Miguel Ángel Canela • Inés Alegre •
Alberto Ibarra

Quantitative Methods for Management

A Practical Approach

Springer

Miguel Ángel Canela
IESE Business School
Barcelona, Spain

Inés Alegre
IESE Business School
Barcelona, Spain

Alberto Ibarra
IPADE Business School
Mexico City, Mexico

ISBN 978-3-030-17553-5 ISBN 978-3-030-17554-2 (eBook)
https://doi.org/10.1007/978-3-030-17554-2

This Springer imprint is published by the registered company Springer Nature Switzerland AG
The registered company address is: Gewerbestrasse 11, 6330 Cham, Switzerland

Preface

More than ever, managers are expected to perform data-driven decision-making in their organizations. We intend to contribute to that trend with a short book which presents some quantitative methods which are frequently used in managerial settings.

Recent developments in the software industry have greatly enlarged the computational capabilities for data analysis purposes. This has facilitated incorporating quantitative methods into all business areas in order to solve interesting questions such as do we have a gender pay-gap in our firm? Can we predict which borrowers will not pay back a loan? What is the optimal mix of fixed and variable salary to increase productivity? What type of promotions enhances customer retention? What is the better route for delivering our products in rural areas?

This is a business-oriented book with a focus on real applications. So, many topics which are typically discussed in statistics textbooks have been skipped. It has 13 chapters, intended to be small doses of quantitative methodology. Every chapter contains a brief discussion of a methodological issue followed by a fully worked business example.

The book is based on our experience in teaching Quantitative Methods at IESE and IPADE. Ingrained in the teaching practice of these business schools is the learning-by-doing philosophy. So, instead of expecting the reader to thoroughly study a method before applying it, we encourage him/her to practice with the example after reading superficially the theory, coming back later for a better understanding.

Also, the examples are not discussed exhaustively, since we wish to provide a level of analysis that is sufficient to understand the topic for a nonexpert (no previous knowledge is required). A professor using this book for a course may take advantage of this by expanding in class what we have intentionally omitted in each chapter.

The book has four parts:

1. The first part uses two interesting examples for a warm-up on summary statistics and distributions.
2. The second part covers the essentials of regression analysis. This is the core of the course, since most business applications of quantitative methods are regression analyses, sometimes in disguise.

3. The third part deals with classification models. This is an atypical topic in textbooks at this level, but we think that the importance of this type of predictive models in today's business justifies devoting some space to introduce them and to discuss how they can be evaluated and validated.
4. The last part of the course covers some elementary methods for time series analysis, with an emphasis on sales forecasting.

The book is complemented with a collection of materials which can be downloaded from a dedicated GitHub site, www.github.com/quants-book. For each example, we provide an Excel file with the data and the calculations and charts involved in the analysis. The data used in some of the examples have been slightly modified, to anonymize their real actors and organizations. Nevertheless, we have tried to maintain the relevance of the setting and the managerial issue to solve.

Note that, from the book's perspective, Excel is just the tool that you have at hand. Readers familiar with a statistical package can use it instead of Excel. Microsoft has developed 13 different versions of Excel since 1987 (and will develop more). So, the reader may have to modify a few details, depending on the operating system and the Excel version, in order to replicate our analyses. Also, we expect readers whose computers have different regional configurations. This may shift how dates and numbers are presented.

The Excel files provided for the examples of this book are in version 97–2003 (extension `xls`). The instructions given for obtaining in Excel the chart sheets included in these files are based on version 2017. We use Excel according to the US conventions, that is, with the full stop (.) as the decimal separator and the comma (,) as the column separator. In Excel functions, the arguments are separated by commas.

For the regression and correlation analyses, we use the `Analysis ToolPak` add-in, which is available for free in the current Excel versions for both Windows and Macintosh. When the add-in is active, the Data tab shows, on the right, the `Data Analysis` button, which calls the corresponding dialog box. If this were not the case, the user would have to activate it. To do that, the instructions provided by the Help utility of Excel could be useful. Since the activation is carried out in different ways in Macintosh and Windows computers, we omit the details, to keep it short.

Some readers may encounter small deviations between their calculations and what is either in the book or in the Excel files. A reason for those minor differences may come from rounding and truncating. In most cases, we have rounded to two decimals the results reported in the book page.

The book includes an appendix with R code for the analysis of all the examples. When providing the code, we have assumed that the potential reader of the appendix is familiar with the R language. Please be careful when copy-pasting the code from an electronic version (PDF) of the book, which may introduce undesired special characters. The code is also available in the GitHub site.

Also in the GitHub site, the interested reader will find all the data sets in CSV format, which is easier to manage in R. In the code provided in the Appendix, the data are always imported from the GitHub remote source. All the figures included in the book have been produced with R. In the chart sheets included in the Excel files, we have tried to get as close as possible to the figures that appear in the book.

Finally, we wish to conclude this preface with two messages. First, none of the methods presented in this book are groundbreaking in any way. All these methods are well known and widely accepted by scholars and practitioners. Based on our collective research, readings, class experience, and professional practice, we have written these chapters with the intention to help you to make informed decisions. Second, we invite you to read this book with patience. Although mastering completely its contents may require years of study and practice, learning from it requires a much lesser effort. Do not be overwhelmed by the apparent difficulty of the subject and take a hands-on approach in trying to learn more about quantitative methods for managers.

Barcelona, Spain — Miguel Ángel Canela
Barcelona, Spain — Inés Alegre
Mexico City, Mexico — Alberto Ibarra

Contents

Part I
Basics

Summary Statistics 1

This chapter deals with summary statistics. The main three summary statistics are the mean, the standard deviation, and the correlation. The first two are discussed in this chapter. The correlation is left for Chap. 3.

The example, based on financial data, shows how to interpret the mean and the standard deviation when the distribution of the data follows a pattern called the normal distribution, which can be described by a bell-shaped curve. Chapter 2 gives more detail on the normal distribution.

1.1 Summary Statistics

Summary statistics are used to explore and/or describe the data. They are useful as far as you and your potential audience are able to interpret them, which can be more or less easy, depending on the data you are dealing with and your experience with the phenomenon you are studying. This is worth to keep in mind because we frequently find reports crowded with statistics for which not even the author can provide an interpretation.

Of the two summary statistics discussed in this chapter, the mean is easier to interpret, although it may not be so when your data have a weird distribution. The standard deviation is more difficult, and it is only useful when the data present a good distribution. We also include in this chapter a comment on percentiles, which are useful with certain types of data.

M. Á. Canela et al., *Quantitative Methods for Management*,
https://doi.org/10.1007/978-3-030-17554-2_1

1.2 The Mean

Let us suppose that we have a set of observations $x_1, x_2, \ldots, x_n$ of a variable X. The number of observations n is called the **sample size**. The **mean**, or average, of these observations is given by the formula

$$\bar{x} = \frac{x_1 + \cdots + x_n}{n}.$$

We denote the mean by $\bar{x}$ (read as x-bar). The mean is taken as a central value, although it is not always so, as shown by the following supersimple example. Take $x_1 = 1$, $x_2 = 2$, $x_3 = 3$ and $x_4 = 10$. The mean is

$$\bar{x} = \frac{1 + 2 + 3 + 10}{4} = 4,$$

which can be obtained in Excel with the formula `AVERAGE(1,2,3,10)`. In this case, the mean does not split the data into two halves, but it leaves 3/4 of the observations on the left and 1/4 on the right. Since this example is so simple, it can be easily guessed what does the trick: that we have piled up three observations on one side, placing the fourth one far away from these three. Describing this more technically, we would say that the distribution of these data around the mean is not **symmetric**. We will come back to this point in the next chapter.

The mean is sometimes called **expected value**, as in the example of this chapter. This terminology may be confusing, since the expected value is not what we "expect" to observe, but the average of what we have already observed. For instance, in a population, the expected value of the number of children per female inhabitant may be 1.7, but no woman is expected to have 1.7 children.

1.3 Median and Percentiles

As said above, the mean is not always the value "in the middle", so that one half of the data points are above the mean, while the other half are below the mean. The statistic with this property is called the **median**.

To calculate the median, we sort the data, so we have $x_1 \leq x_2 \leq \cdots \leq x_n$. If n is odd, the median is equal to the data point in the middle of the sorted list, that is, in place $(n+1)/2$. If n is even, the median is the midpoint of the data points in places $n/2$ and $(n/2)+1$. Note that, in this case, any value between the two central points splits the data into two halves. We take the midpoint as the median to get an operational definition.

In the example above, the median is 2.5, far from the mean 4, which illustrates the lack of symmetry. Statisticians typically look at the gap between the mean and the median to evaluate the symmetry of the data.

50% of the observations are equal to or less than the median. Replacing the 50% by any other percentage, we get the definition of a **percentile**. For instance, the 25% percentile leaves one-fourth of the data on the left and three-fourths on the right. Although this is exactly true only when the number of observations is a multiple of 4, the percentile notion is still useful and appealing.

1.4 The Standard Deviation

The minimal description of the data should contain a **measure of central tendency**, such as the mean or the median, and a **measure of dispersion**. The latter tells us about how concentrated around the central value the data are. Because of its mathematical properties, the **variance** is the preferred dispersion measure. It is defined as

$$s^2 = \frac{(x_1 - \bar{x})^2 + (x_2 - \bar{x})^2 + \cdots + (x_n - \bar{x})^2}{n - 1}.$$

The **standard deviation** is the square root of the variance. Note that the standard deviation has the same units as the data, but the variance has not. If the data come in dollars, both the mean and the standard deviation are in dollars, but the variance is in squared dollars. So, although variances may appear in many formulas in Statistics, we usually report standard deviations, which are easier to interpret than variances.

The variance is usually denoted by s^2, and the standard deviation by s. Sometimes the Greek version σ (sigma) is used instead of s. For the example above, the Excel formula `STDEV(1,2,3,10)` performs the calculation

$$s = \sqrt{\frac{(1 - 4)^2 + (2 - 4)^2 + (3 - 4)^2 + (10 - 4)^2}{3}} = 4.08.$$

Although standard deviations are frequently reported, not everybody has a clear idea of what the numbers themselves mean. The interpretation of the standard deviation in practical terms is based on a probabilistic model, called the normal distribution, as discussed below.

Note. There is an alternative version of the formula of the variance with n in the denominator instead of $n - 1$. But, for the sample sizes used in business, this point is irrelevant.

1.5 Excel Functions

The Excel functions for the mean, the median, the standard deviation, and the variance are, respectively, `AVERAGE`, `MEDIAN`, `STDEV`, and `VAR`. These functions apply to a range of data, such as in `AVERAGE(B2:B242)`, which calculates the average of the entries in the 241 cells of the range from `B2` and `B242` (this is the range that contains the returns discussed in the example). If a cell included in that range is either empty or has non-numeric data, Excel ignores it.

For the percentiles, we have to specify the proportion of observations lower than the percentile requested. For instance, `PERCENTILE(B2:B242, 0.9)` returns the 90% percentile, so 9/10 of the data are below that statistic and 1/10 are above. In the example of this chapter, the values `PERCENTILE(B2:B242, 0.025)` and `PERCENTILE(B2:B242, 0.975)` are used as the extremes of an interval which contains 95% of the data.

Note. With the best intention, Excel offers many variants of the variance formula, which have been changing across versions (`VAR.P`, `VAR.S`, `VARA`, `VARP`, and `VARPA`), creating a mess. The same can be said for the standard deviation. It also offers two additional variants of the percentile (`PERCENTILE.EXC` and `PERCENTILE.INC`). The differences among these versions are not relevant for managers.

1.6 The Histogram

A **histogram** is a special case of a bar diagram. To draw a histogram, we partition the interval covered by the data into a collection of subintervals of equal length, sometimes called **bins**. Upon each of these intervals, we place a bar whose height is proportional to the number of observations falling within that interval. We use a histogram in the discussion of the example of this chapter.

1.7 The Normal Distribution

Statisticians use a set of carefully chosen curves as theoretical models for histogram profiles. The most popular of these curves is the bell-shaped **normal distribution**. The mathematical properties of this curve are discussed with more detail in Chap. 2. Our analysis in this chapter is mainly visual.

In a normal distribution, the formula $\bar{x} \pm 2s$ gives two limits between which lies (approximately) the 95% of the population. This formula makes the normal distribution easily manageable and provides a direct interpretation of the standard deviation. If the data present a clear departure from the normal distribution, it may not be easy to provide an interpretation for the standard deviation.

1.8 Example: Tata Daily Returns

1.8.1 Presentation

The National Stock Exchange Ltd. (NSE), located in Mumbai, is the leading stock exchange in India. This example is based on historical data on **stock prices** from the NSE, extracted from the Yahoo Finance India website. More specifically, the data set contains daily prices for Tata Auto, covering the year 2014.

Historical stock prices are typically published in the open-high-low-close (OHLC) format. This example uses adjusted closing prices, which are the daily closing prices adjusted for all applicable splits and dividends. The data can be found in the file `tata.xls` (sheet `Prices`, range `B2:B243`).

For most investors, what matters is not the price itself, but how the price changes, since this gives the benefit or loss of investment. This change is tracked through the returns. **Daily returns** can be calculated in more than one way. The simplest approach is that of the simple return, defined as

$$\text{Return}(t) = \frac{\text{Price}(t) - \text{Price}(t-1)}{\text{Price}(t-1)} = \frac{\text{Price}(t)}{\text{Price}(t-1)} - 1.$$

Here, t stands for the day ($t = 1, 2, 3, \ldots$). Since returns are proportions, they are usually expressed as percentages. For instance, the price of Tata Auto stock on January 2nd is 366.8145, falling down to 357.3076 on January 3rd. So, the return on January 3rd is

$$\frac{357.3076}{366.8145} - 1 = -2.59\%.$$

Note. Hand and pencil, the calculation above will give you (in absolute value) 0.0259, so you have to multiply the result by 100 to get the percentage. In Excel, 0.0259 and 2.59% are the same things in different formats. Selecting a range of cells in Excel, Ctrl+1 (in Windows) or Cmd+1 (in Macintosh) displays the `Format Cells` dialog box, which allows format changes.

The returns have been calculated in the sheet `Returns`. Note that, since two consecutive prices are involved in the calculation of a daily return, the series of returns has to be one day shorter than that of prices, starting on day January 3rd (range `B2:B242`).

1.8.2 Plotting the Data

Figure 1.1 is a line plot for the return of the adjusted closing prices of Tata Auto stock. We see here a standard pattern of up-down variation, with no evident trends nor volatility episodes. This means that, when looking for a pattern for the distribution of the returns, it can be useful to ignore the time. In that case, a histogram becomes useful.

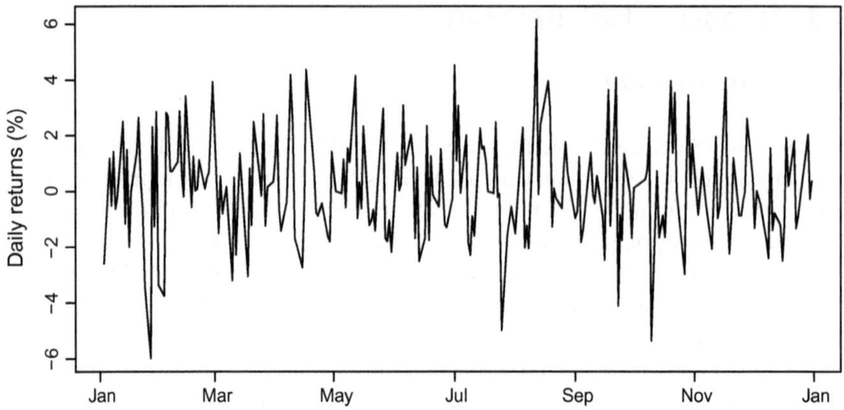

Fig. 1.1 Tata Auto daily returns (line plot)

The sheet `Line plot` of the Excel file contains an Excel version of Fig. 1.1. Such a line plot can be obtained as follows:

- We select the range containing the data (sheet `Returns`, range `B2:B242`).
- We click the tab `Insert`.
- We click on the corresponding icon in the group `All charts`.
- In the first row of options (`2-D line`), we click the first one (`Line`).
- The chart, unedited, will appear in the middle of the screen. We can move it within the sheet that contains the data, or put it in a separate sheet, by right-clicking in the chart and selecting `Move Chart`.

By default, Excel uses, as labels for the horizontal axis, indexes 1, 2, etc. In Fig. 1.1, we have used a date scale that looks more appealing. In the Excel file, we have replaced the indexes by the months, as yyyy-mm. To do that, you can proceed as follows:

- Right-click on the chart and select `Select Data Source`. The corresponding dialog box appears.
- In the box, click on `Edit` in `Horizontal (Category) Axis Labels`. The dialog box `Axis Labels` appears.
- Enter the range of dates, as `=Returns!A2:A242`, and click on `OK`.
- The dates will have replaced the indexes in the horizontal axis, but they may not be in the right format. You can polish this by right-clicking on the axis and doing `Format Axis » Number » Type yyyy-mm`.

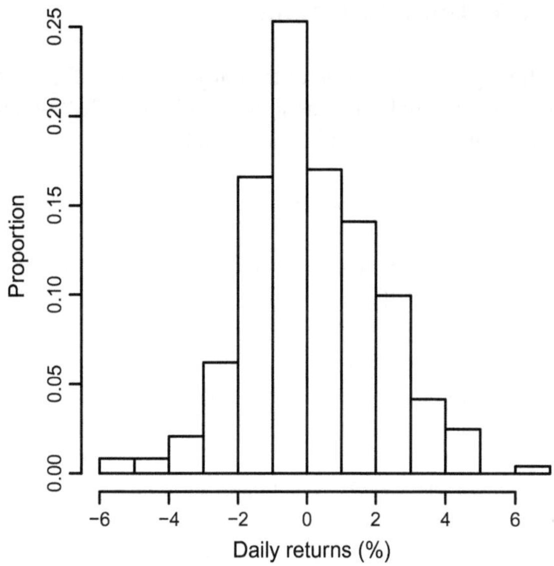

Fig. 1.2 Tata Auto daily returns (histogram)

Figure 1.2 is a histogram of the daily returns. In Excel, we can get a similar graphical representation. The `Analysis ToolPak` provides a bar plot that can be edited to look as the histogram of Fig. 1.2. The process is as follows:

- Install `Analysis ToolPak`.
- In the tab `Data`, click on `Data Analysis`.
- In the dialog box that appears, click `Histogram` and then `OK`.
- In the dialog box that follows, enter in the box `Input Range` the range covered by the data (`B2:B242`). This will produce a new sheet with a table of counts for collections of bins.
- Create a bar chart based on this table. The width of the bars can be augmented so the chart looks like a typical histogram.

1.8.3 Statistics

Some useful statistics associated with daily returns are the mean and the standard deviation. In this context, the mean is called the **expected return**. In the sheet `Returns`, the formula `AVERAGE(B2:B242)` gives us an expected return of 0.0014 (0.14%). In finance, the variability of the returns is typically associated with the risk. So, the standard deviation of the returns can be used as a measure of risk. With the formula `STDEV(B2:B242)`, we get 0.0188 (1.88%).

1.8.4 The 95% Interval

Although the histogram of the daily returns of Tata Auto presents a certain departure from the normal curve, as shown in Fig. 1.3, the "2-sigma" interval works reasonably well. The limits given by the formula are

$$0.14\% - (2 \times 1.88\%) = -3.63\%, \qquad 0.14\% + (2 \times 1.88\%) = 3.90\%.$$

This is a theoretical calculation. 95% of the observations are expected to fall within these limits if the data follow a normal distribution pattern. The 2.5 and 97.5% percentiles are −3.37 and 4.08%, respectively, not far from these theoretical values.

We can assess how good are the limits given by the formulas of the normal distribution in a naive way as follows. Among the 241 returns, 15 (6.2%) fall out of the interval defined by these formulas: 5 on the left (days 01–27, 02–03, 07–25, 09–23 and 10–10) and 10 on the right (days 02–28, 04–09, 04–17, 05–12, 07–01, 08–12, 08–18, 09–22, 10–20 and 11–17). So, we are very close to 95%.

Note. The exact value which must be used to get the 95% interval is 1.96. Rounding it allows for mental calculation and facilitates memorization.

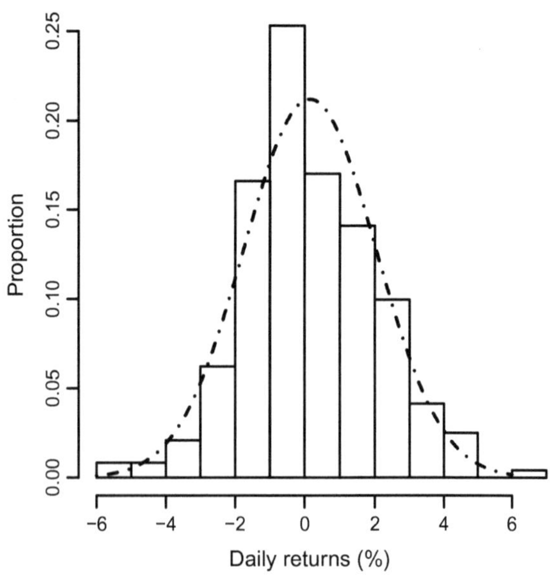

Fig. 1.3 Matching the normal distribution

1.9 Useful Tips

- Knowing your data is important. Summary statistics and graphs help you to understand the data before starting a more elaborated analysis. *Always perform an exploratory analysis.*
- A visual display of the data (e.g., a line plot or a histogram as those used in the example) can be very helpful. A simple descriptive statistics table may not convey quickly the story behind a phenomenon in the way a graph can do. Data visualization is important, both for the analyst and for the recipients of the analysis. When possible, *develop a visual representation of the data.*
- Absolute (e.g., average) and relative (e.g., percent change) figures may be misleading when observed in isolation. Government expenditures look very large in the news, since government budgets are much larger than ours. Relative figures (e.g., growth of a business unit) are not easy to interpret without a reference. It is not only important to know the profits of a firm in this quarter but also how that profit compares against the same figure from the previous year. *Pay attention to both absolute and relative figures.*
- *Never guess the distribution of data from a measure of central tendency such as the mean or the median.* A few months ago, a local politician suggested that people should use more frequently the bicycle, since most citizens lived on average 5 km away from their work. That statistic would still be true if one citizen lives 500 meters away from the office and another citizen 9.5 km away.
- Measures of central tendency should be complemented with a measure of dispersion. The standard deviation is the best candidate when the data have a bell-shaped distribution. On the other side, it can be useless for very skewed distributions, as we will see in Chap. 2. In that case, percentiles are more informative.
- Mean, average, and expected value are three ways to refer to the same thing. The mean is a reliable and useful statistic when our data have a symmetric distribution. But taking the mean as representative of a sample can be dangerous when the data have a skewed distribution. Imagine you live in a middle-class town. The average annual salary of the people in the town is $30,000. If suddenly Mark Zuckerberg decides to move into your town, the average salary will dramatically increase, but it will not be informative of the purchasing power of the town's people. In such a context, the median would be more useful. *Replace the mean by the median for variables with a skewed distribution.*

Probability Distributions

2

This chapter is a very elementary introduction to probability distributions. We first introduce the concept of probability in managerial terms and then explain how to operationalize the probability distribution of categorical and continuous variables.

The case of the normal distribution is discussed with more detail. It is shown in the example, based on Twitter data, how a variable may look "more normal" after applying a transformation such as the logarithm.

2.1 Probability

Many business decisions are taken based on an estimation of how likely an event is to occur (e.g., if a borrower will repay a loan); however, more should be done to update models that inform decision-making processes. We do little effort in tracking the success or failure of our past estimations; consequently, we limit our learning. The **probability** of an event is a number which indicates how likely that event is to occur. The minimum probability is 0 for an event that it is not expected to occur whatsoever. The maximum probability is 1 for an event that we expect for sure.

Probability is mainly classified into two domains: objective and subjective. The former is the one in which the likelihood of an event can be estimated regardless of a personal opinion. The latter is derived from personal judgment and it can be based on experience and interpretation of the available information. For management, it is important to understand both, since objective probabilities can provide managers with a better understanding of the possible outcomes in their decision-making processes, and subjective probabilities allow to improve models by updating relevant information that is not included in a calculation.

The subjective approach is not easy. Assessing probabilities in real life turns out to be difficult and not very intuitive. Several studies have shown that humans are not

M. Á. Canela et al., *Quantitative Methods for Management*,
https://doi.org/10.1007/978-3-030-17554-2_2

very good calculating probabilities and that biases have a strong effect on probability judgments. This course follows an objective approach, based on data analysis.

Daniel Kahneman (a Nobel Prize laureate) made a great contribution with his "law of the small numbers" which dictates that people tend to make inferences about a population based on the results of a small sample. When thinking about probabilities in management, we have to be aware of how our assessments can be easily influenced by generalizing ideas based on small samples.

2.2 Probability Distributions

Statisticians are interested in the probabilities of the different outcomes of a variable. These probabilities can be managed in two different ways, depending on the variable being categorical or continuous. Let us start with a **categorical variable**, that is, a variable which can only take a finite collection of values. Two simple examples are as follows:

- The gender of a newborn can take two values, MALE and FEMALE. Let us suppose that the probability of female newborn is 0.48, and that of a male baby is 0.52. This means, in practice, that, for a large sample of births, we expect (approximately) 48% of girls and 52% of boys.
- The outcome of a die can take values 1, 2, 3, 4, 5, and 6, all with the same probability, 1/6. This means that, if we toss the die thousands of times, all the outcomes will occur in a similar proportion.

By assigning a probability to every value of a categorical variable, we get the **probability distribution** of that variable. We call this probability distribution a **discrete distribution**. Since there is just a list of possible values, we have a simple way to specify their probabilities. The case of a continuous variable is more complex, and needs a bit of mathematical apparatus, as we see in the next section.

Figure 2.1 is the visualization of a discrete probability distribution. Suppose that we toss two dice and take the mean. The variable whose distribution we see in the figure is the result of that calculation. There are 11 possible values for this variable (1, 1.5, 2, 2.5, …, 5.5, 6), whose probabilities are easily calculated.

For instance, the most probable value is 3.5 (the highest spike in Fig. 2.1), which can be obtained as the mean of 1 and 6, of 2 and 5, of 3 and 4, of 4 and 3, of 5 and 2, and of 6 and 1. In total, 6 cases, out of the 36 possible combinations that we can get with the two dice. So, the probability is 1/6. The probabilities of the other cases are obtained with similar arguments.

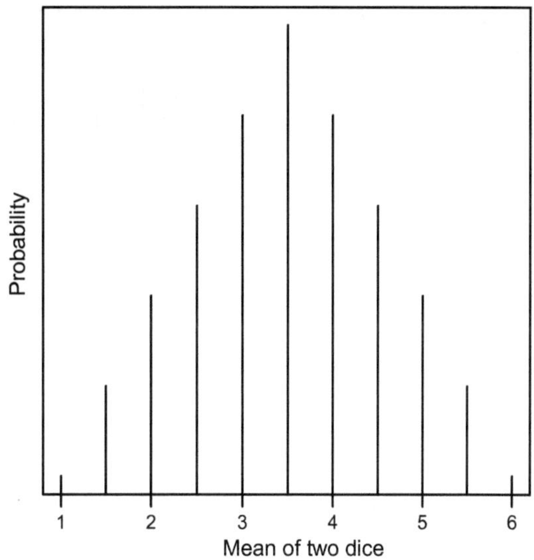

Fig. 2.1 Discrete probability distribution

The distribution of Fig. 2.1 is a **symmetric probability distribution.** In the center, we have the most probable value, and the probability decreases on both sides in the same way. The symmetry of the distribution will be examined in this chapter in the different data sets that appear successively.

2.3 Continuous Probability Distributions

A **continuous variable** is one that, for any pair of values, can take every intermediate value. For a continuous variable, we do not have a finite collection of potential values, so we cannot assign a probability to each value. For continuous variables, we look at the probabilities of intervals. In this context, the typical question would be: given two values x_1 and x_2 of a continuous variable X, what is the probability of X falling between x_1 and x_2 (i.e., of $x_1 < X < x_2$)?

A **continuous distribution** is managed through a **probability density curve**, like that of Fig. 2.2. The probability of an interval is calculated as the area under the density curve for that interval. This involves a mathematical technique called **integral calculus.** We do not give details here about that, but a visual inspection of a graphical representation like Fig. 2.2 will give you a rough idea about the distribution represented there.

What do we see in this figure? First that the total area under the curve must be 1, which is the total probability distributed. Also, an interval containing only negative values has probability zero. Third, the peak is about 2.5, so the variable must take values more often in that part than in anywhere else.

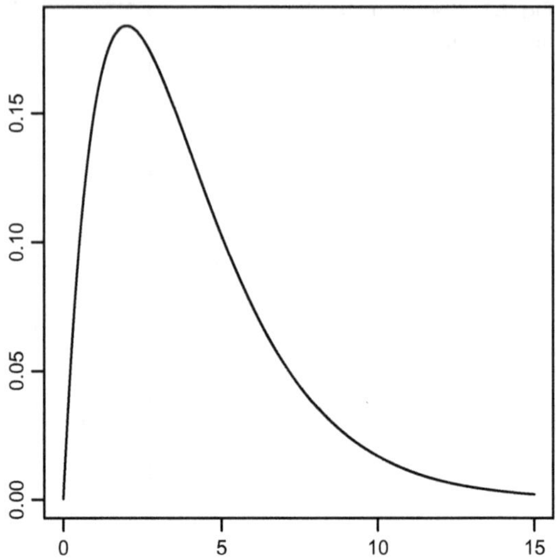

Fig. 2.2 Probability density curve

This distribution of Fig. 2.2 is not symmetric, in contrast with that of Fig. 2.1. We call it a **skewed distribution**. Note that, in contrast with the normal distribution already mentioned in Chap. 1, we have a right **tail**, but no left tail. For skewed distributions, statistical formulas based on means and standard deviations are useless, as discussed in the example of this chapter. Skewed distributions are not so rare. For instance, distributions of income in various contexts are typically skewed.

Note. In Fig. 2.2, and in Fig. 2.3, the probabilities are the areas below the curve, not the numbers indicated in the vertical axis.

2.4 The Normal Distribution

Statistical methods typically assume that the distribution of the probability across the different outcomes of the variable subject to the analysis follows a certain pattern which has nice properties. There are many potential models for a probability distribution in Statistics textbooks. For continuous data, statisticians' favorite model is the **normal distribution**, which has already appeared in the preceding chapter of this book.

Figure 2.3 shows three examples of normal probability curves. Let us start with the curve drawn with a solid line, called the **standard normal**. Even if this is a beautiful, well-known curve, the equation may look a bit scary,

$$y = \frac{\exp(-x^2/2)}{\sqrt{2\pi}},$$

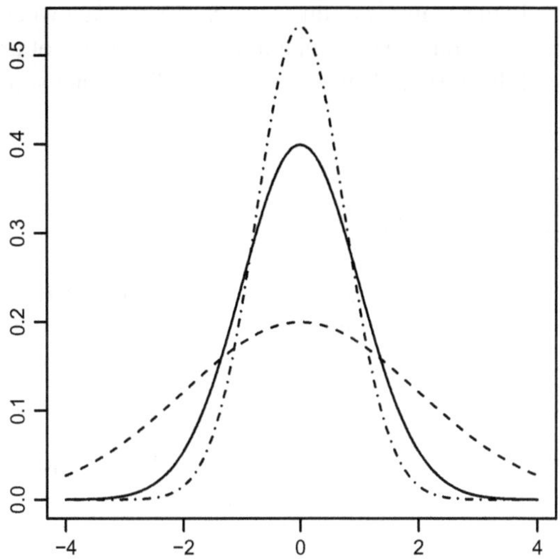

Fig. 2.3 Normal probability curves

so we manage everything through the computer, which knows the normal distribution pretty well.

How do we distinguish among the different normal probability curves? By using two parameters, which are the **mean** and the **standard deviation**. For instance, the standard normal distribution has zero mean and unit standard deviation.

The mean of the distribution is understood as the limit of the mean value obtained in a sample of many independent observations, when the number of observations tends to infinity. In practice, this means that, for big samples, we can trust the mean of the data as an estimate of the mean of the distribution. The standard deviation can be understood in a similar way.

Keeping the standard deviation equal to 1, any other choice of the mean will lead to the same shape, but not centered at $x = 0$. With a different choice for the standard deviation, the curve looks equally bell-shaped, but flatter (for $s > 1$) or more peaked (for $s < 1$). The two dashed curves of Fig. 2.3 have been obtained setting $s = 2$ and $s = 0.75$, respectively.

What makes the normal distribution so interesting? That we have a simple formula for making predictions, based on the mean and the standard deviation. More specifically, if a variable has a normal distribution, the probability that it falls between the limits $\bar{x} - 2s$ and $\bar{x} + 2s$ is (approximately) 95%. This has been illustrated in the example of Chap. 1. Even if the distribution was not normal, it was reasonably symmetric and had tails on both sides. The formula of the 95% interval of the normal distribution gave us an approximation which was still useful for practical purposes.

Nevertheless, if we apply this formula to a skewed distribution like that of Fig. 2.2, part of the interval calculated in this way will invade the negative side, which does not make sense for a variable which cannot take negative values.

In the example of this chapter, we use data whose distribution is very far from the normal, much worse than in Fig. 2.2. This is not unfrequent in real data. Even if we call the normal distribution "normal", it is not something that happens "normally".

2.5 Transformations

Sometimes, the statistical analysis is not performed on the raw data, but on data resulting from a transformation. The transformation may facilitate the analysis, for instance changing the original distribution into one which is closer to the normal pattern.

The transformation is based on a mathematical function. The **logarithmic transformation** is the archetypical example. It is based on the natural logarithm function, which, in Excel, is `LN`. This function is the inverse of the exponential function `EXP`, meaning that `y = LN(x)` is equivalent to `EXP(y) = x`. The exponential function returns the powers of the number $e = 2.718282$.

Let us illustrate this with a numerical example. Take $x = 2$. Then

```
EXP(2) = 7.389056
```

is the square of the number e. Now, the natural logarithm of this number gives us the original x value,

```
LN(7.389056) = 2.
```

Note. These numbers have been rounded to six decimals. The number e has infinitely many decimals.

Now, coming back to Statistics, let us suppose that X is a variable which comes in our raw data set, such as the annual salary in euros (which could have a distribution like that of Fig. 2.2). Then, the variable defined as $Y = \ln(X)$ would be the transformed variable.

What is the benefit of such a transformation? The distribution of the transformed variable can be closer to a desired pattern like the normal distribution. This is not the only motivation for a transformation in a statistical analysis, but it is the one discussed in the example of this chapter.

Note that, since the exponential of any number is positive, only positive numbers have a logarithm. So, the only variables to which a logarithmic transformation can be applied are those who only take positive values.

2.6 Time Series Data

As we said in the Chap. 1, we look at the probability curve as a theoretical model for the histogram. This type of analysis only makes full sense for data which come all from the same source and have been obtained under the same conditions.

The typical example occurs when we extract a random sample from a big population. The histogram, and the type of probability curve that we may see behind it, helps us to understand how a continuous variable is distributed across the individuals of that population.

But this is not, in general, appropriate for data collected at regular intervals of time (e.g., daily, or hourly). We call this type of data **time series data**. Part 4 of this book is devoted to the analysis of that type of data.

Time series data are preferably represented using a line plot. The plot may reveal that the very concept of the probability distribution does not apply to those data. This is the case when the line plot is not flat (e.g., when there is an upward trend), so that we see that the mean value changes as time runs.

When the data are obtained from a sample of a population, the histogram can be used for describing not only the sample, but also the population, predicting how the values of the variable observed will be distributed for the population units not included in the sample. But, with time series data, you have to be more careful if you wish to predict future values of the variable, because the distribution of the actual data cannot, in general, be assumed to be valid in the future (we did that in the example of Chap. 1, based on our exploratory analysis). Then the histogram describes only the actual data. The example of this chapter illustrates this point.

2.7 Example: The EuroLeague Final Four

2.7.1 Presentation

The EuroLeague is the European-wide-top-tier basketball club competition that is organized by EuroLeague Basketball, since 2000, for eligible European basketball clubs. It has the highest attendance among professional indoor sports leagues outside of the United States.

The EuroLeague Final Four, which closes the EuroLeague, took place in 2017 in Istambul. The four contenders were Olympiacos Piraeus, Real Madrid, CSKA Moscow, and Fenerbahce Istanbul. The semifinals took place on Friday, May 19, Olympiacos winning over CSKA and Fenerbahce over Real Madrid. Two days later, CSKA won the third place (94–70) and Fenerbahce defeated Olympiacos (80–64) in the final.

The managers at the EuroLeague Basketball offices in Barcelona were interested in learning about the Twitter activity generated by the EuroLeague events. So, they asked a freelance data scientist to collect all the tweets posted during 5 days, from Thursday 18 to Monday 22, in several languages (we only use the tweets written in Turkish in this example), and to prepare a statistical report based on the data collected.

2.7.2 The Data

Twitter makes data available through several API's (Application Programming Interfaces). In particular, the Twitter Search API returns a collection of relevant tweets matching a specified query. The data scientist uses this API to extract all the public tweets in Turkish, containing the hashtag "#fener4glory", posted on May 18–22. A total of 643,433 tweets were captured.

The file `euroleague.xls` contains the tweet counts, by minute (sheet `Raw data`) and by hour (sheet `Aggregate data`). We start the raw data analysis, using the counts by minute of the sheet `Raw data`.

2.7.3 Raw Data Analysis

Figure 2.4 is the line plot for the raw data. The data look a bit crazy, with two hot periods when the Fenerbahce matches were played. The tweet production at those times was so intense that this fact completely dominates the picture. This type of behavior is not rare in data on social network activity (do not expect normal distributions in that context).

Since these are time series data, the histogram may only be useful for a description of the period covered by the data. But this is not the case here, as shown in Fig. 2.5. In addition to the fact that the histogram hides the within-days variation pattern that human activity typically shows and the between-days variation pattern specific of these data, it is not clear what kind of model we may choose for this distribution.

The mean and the standard deviation are easily obtained in Excel,

```
AVERAGE(B2:B7201) = 89.37,

STDEV(B2:B7201) = 369.01,
```

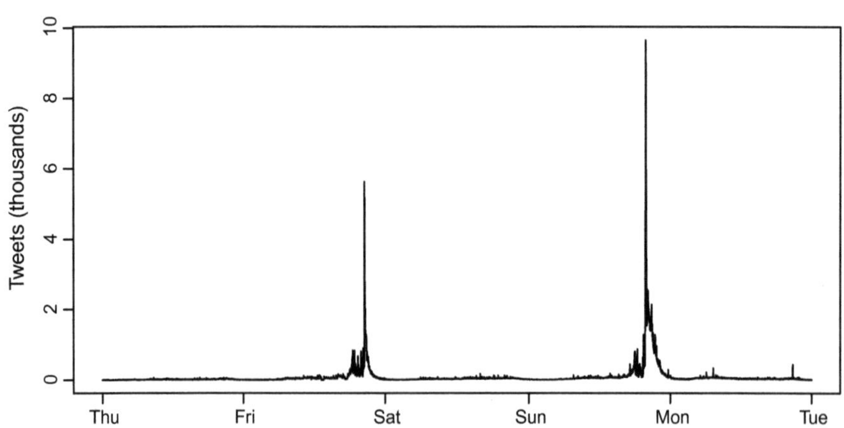

Fig. 2.4 Tweets per minute (line plot)

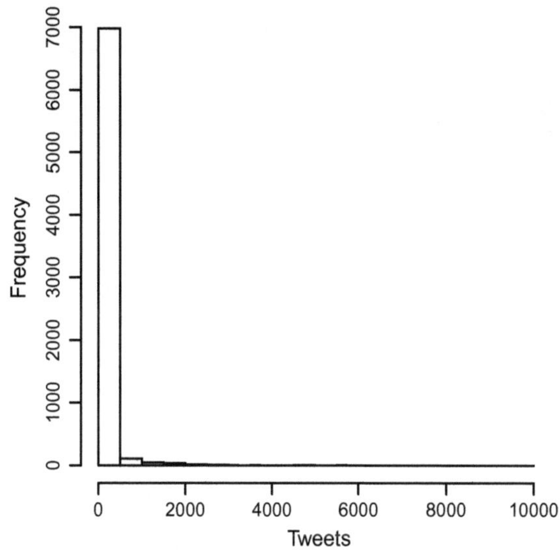

Fig. 2.5 Tweets per minute (histogram)

respectively. These two statistics are quite useless, specially the standard deviation. The formula of the 95% limits for the normal distribution does not make sense here, since the interval given by the formula would invade the negative half-line.

More interesting would be to get a few percentiles. The median

```
MEDIAN(B2:B7201) = 28
```

tells us that, in about one half of the minutes sampled, less than 28 tweets were posted (in the period covered by the data). The 25 and 75% percentiles,

```
PERCENTILE(B2:B7201, 0.25) = 12,

PERCENTILE(B2:B7201, 0.75) = 54,
```

are also informative.

2.7.4 Aggregate Data Analysis

We take a look now at the hourly data of the sheet `Aggregate data`. Figure 2.6 shows the corresponding line plot. Even if a lot of detail has been lost, a within-day pattern of variation can be guessed from this plot.

Does the distribution become "more normal" when we aggregate the data? This idea, which some people share, is not supported by the histogram of Fig. 2.7. But it is true that, as in the case of the line plot, the visualization of the data is more informative now.

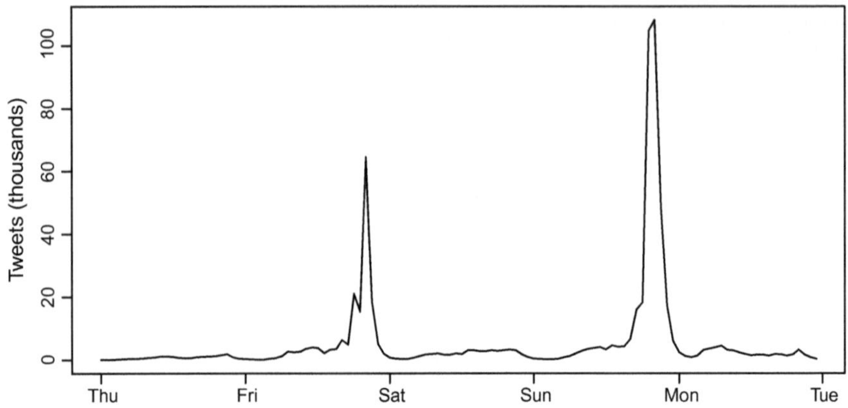

Fig. 2.6 Tweets per hour (line plot)

Fig. 2.7 Tweets per hour (histogram)

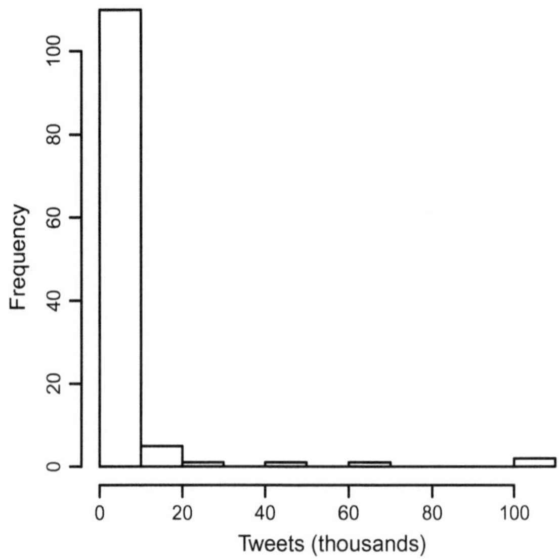

The mean and the standard deviation of the aggregate data are 5,361.94 and 15,403.08, respectively. The comments would be the same as those given for the raw data, although the percentiles would be more informative in this case. The median is 1,756.5, telling us that, half of the time, the hourly counts have been above this number.

This example helps us to see that, in spite of the popularity of the normal distribution, it is not safe to take for granted that the data that we collect will follow that pattern, and also that it is not merely a question of aggregation.

2.7.5 Logarithmic Transformation

Does the distribution get nicer with a logarithmic transformation of the data? Statisticians tell us that the logarithmic transformation can make a twist on a skewed distribution. To illustrate this point with the EuroLeague aggregate Twitter counts, we apply the function `LN` to each of the cells in the range `B2:B121` of the sheet `Aggregate data`.

The line plot is Fig. 2.8. The daily pattern is quite clear, though it is distorted by Final Four events. Of course, these patterns are country-specific. The vertical scale, even if it is logarithmic is not that difficult to understand. Owing to the properties of the logarithm function, the scale is multiplicative, meaning that going up one unit means multiplying by the factor $\exp(1) = 2.718$ (the famous number e).

The mean and the standard deviation of the log counts are 7.43 and 1.35, respectively. The histogram is shown in Fig. 2.9. Here, we have changed the vertical scale from frequency to proportion, in order to match the scale of the normal probability curve that has been superimposed (whose mean and standard deviation are those extracted from the data). The fit is not perfect, of course, but it is not a poor job.

How good is a 95% interval based on the mean and the standard deviation of the log-transformed data? The limits are

$$7.43 - (2 \times 1.35) = 4.74, \qquad 7.43 + (2 \times 1.35) = 10.12.$$

There are 114 observations within the limits, exactly the 95% of the sample. So, we conclude that the log transformation does the job in this case, helping to "normalize" the data.

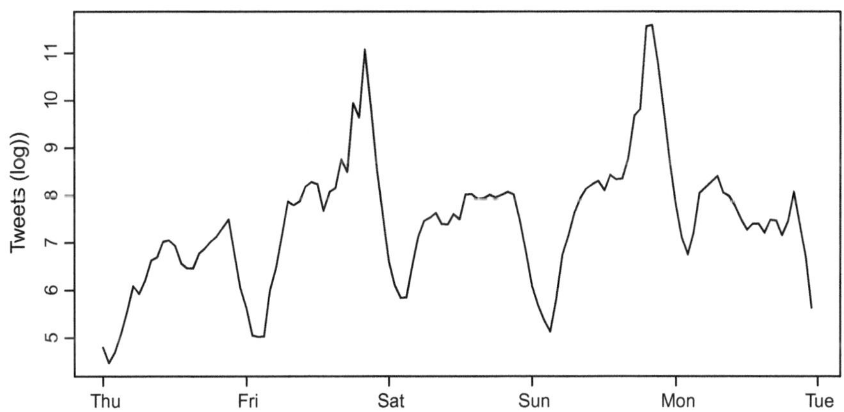

Fig. 2.8 Tweets per hour, log scale (line plot)

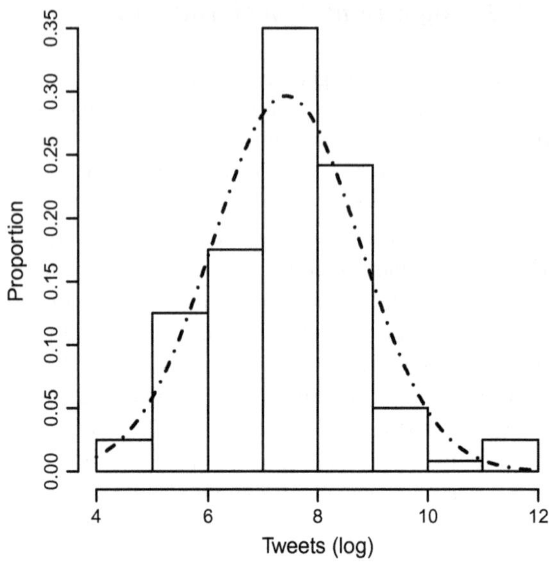

Fig. 2.9 Tweets per hour, log scale (histogram)

2.8 Useful Tips

- Understanding the likelihood of an event is critical in risk management and, consequently, it is a crucial issue for managers. The probability provides a numerical measure of that likelihood. Since uncertainty is everywhere, understanding and assessing probabilities is an important asset. *Dedicate time and effort to estimate probabilities.*
- In a managerial setting, when dealing with uncertain events, probabilities need to be reasonably precise, but no more. For the probability of a client buying our product, or the probability that bad weather affects our sales, the difference between 20 and 21% is irrelevant. *Do not be worried about precision, but about having a rough estimate* of the frequency with which that event will occur, a number that will allow you to make reasonable predictions and plans.
- In real life, very frequently the data does not follow a normal distribution. But the shape of the probability distribution may be guessed from the experience. For instance, in the real estate market, we expect the distribution of house prices to be heavily skewed to the right. *Get familiar with the data and the context and type of business before making assumptions about probability distributions.*
- The probability to get tail while tossing a coin is 50%. But it is perfectly possible to toss a coin four times and get four heads. Theoretical probabilities are only realized when the trial is carried out a large number of times. This is called "the law of large numbers". *Be careful with theoretical probabilities for variables that are observed just a few times.*

Part II
Regression Analysis

The Regression Line

3

The second part of this book is devoted to regression analysis. This chapter presents the main properties of the simplest regression model, the regression line. Chapters from 4 to 7 deal with regression analysis in general.

The example of this chapter uses data on sales and prices to illustrate the calculation and the interpretation of the coefficients of the regression line. The regression equation is used as a model to explain the impact of the price on the sales.

3.1 A Trivial Example

We start with a simple example which may help to grasp the essentials. Let us suppose that we observe the height and the weight in a sample of five 40-year old Italian men (Table 3.1).

Putting the height in the horizontal axis and the weight in the vertical axis, we can plot these observations as the five points of Fig. 3.1. This representation is called a **scatter plot**. Scatter plots are typically used to explore the association between two variables. We will see later in this chapter how to manage them in Excel.

The scatter plot suggests a positive relationship between height and weight, meaning that, the more height, the more weight. The **regression line**, which is also included in the figure, is the line that fits these points in the best possible way.

The regression line can be expressed, in mathematical terms, as an equation of the form $y = a + bx$, called the **regression equation**. In Fig. 3.1, $a = -84.47$ and $b = 0.94$. So, the regression equation for the data of Table 3.1 is

$$\text{Weight} = -84.47 + 0.94\ \text{Height}.$$

We look at this equation as a predictive model. Entering a height value in the equation, we obtain the corresponding weight value. We call this calculated weight

M. Á. Canela et al., *Quantitative Methods for Management*,
https://doi.org/10.1007/978-3-030-17554-2_3

Table 3.1 Height and weight of five men

Height (cm)	185	179	192	187	182
Weight (kg)	87	82	95	93	89

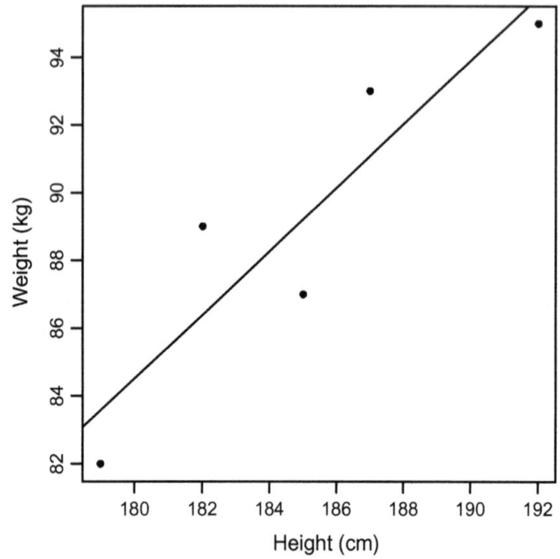

Fig. 3.1 Regression line, weight versus height

a **predicted value**, to indicate that it is not the real weight, but the outcome of a predictive model.

For instance, for a height of 185 cm, observed in the first individual of our sample, we get a predicted weight of 89.2 kg. This is not the same as the actual weight of this individual, which is 87 kg, so our prediction has an error:

$$\begin{aligned}\text{Prediction error} &= \text{Actual weight} - \text{Predicted weight}\\ &= 87\text{ kg} - 89.2\text{ kg}\\ &= -2.2\text{ kg}.\end{aligned}$$

The smaller the prediction errors, the better the model fits the data. In the graphical representation, this can be expressed in terms of the closeness of the points to the regression line. The statisticians use the **correlation**, denoted by R, to measure this. In this case, $R = 0.908$, which provides two pieces of information:

- R is positive, meaning that the association of height and weight is positive (the taller, the weightier), as seen in Fig. 3.1.
- The absolute value of R is close to 1, meaning that the predictions of our model are accurate, with small errors.

3.2 Simple Linear Regression

Loosely speaking, **simple linear regression** is the process by which, given n data points $(x_1, y_1), (x_2, y_2), \ldots, (x_n, y_n)$, we obtain the equation that fits these points in the best possible way. We call it the regression equation.

The regression equation can be represented as a line in an XY plane. Figure 3.1 illustrates this. That line is the regression line. Because of the graphical representation of the regression equation, its coefficients receive special names: a is the **intercept** and b is the **slope**.

The regression coefficients are calculated with formulas which are not too difficult, and can be found in any textbook (or in the online Excel help center). These formulas are implemented in Excel functions, discussed later in this chapter, which are easy to manage. So we skip here the formulas, sticking the discussion to the Excel functions. We regard the calculation of the coefficients as the execution of an **algorithm**. This way of thinking is useful when we consider more complex algorithms, such as those provided by **machine learning** methods, for which the formulas are not accessible to people without mathematical training.

Note. Machine learning is a branch of artificial intelligence concerned with developing models for prediction and other purposes, based on data.

The statisticians regard the regression line under the following perspective. The data points are taken as the observations of two variables, X and Y, on a sample. The variable that we put on the left side of the equation (Y) is called the **dependent variable** and, that on the right side (X), the **independent variable**.

Depending on the specific application, the objective of a statistical analysis based on a regression line could be:

- To describe the relationship between Y and X.
- To predict the value of Y for a given value of X.
- To test the effect of X on Y.

3.3 Predicted Values and Residuals

A value of the dependent variable Y calculated with the regression equation is called a predicted value. The prediction error, that is, the difference of the actual value minus the predicted value, is called, in a linear regression context, **residual**. So, we have a decomposition

$$\text{Actual value} = \text{Predicted value} + \text{Residual}.$$

As we said above, the regression line fits the data in the best possible way. More specifically, this means that the regression coefficients are chosen by the regression algorithm in such a way that the sum of the squares of the residuals is minimum. This is called the **method of the least squares**. This method is the choice of statisticians to operationalize the "best fit", which is a subjective notion.

If we trust our sample, we may use a regression equation as a predictive model, to estimate the value of Y from the value of X for an individual not included in the sample. In such applications, we should restrict our predictions to values of X within the range covered by the data. For instance, our equation for predicting the weight should not be applied to a height of 160 cm. That would be a case of **extrapolation**, one of the worst sins in Statistics.

3.4 Interpretating the Regression Coefficients

Although we usually need the intercept in a regression equation to get a good fit, we do not pay much attention to it, since it rarely has any meaningful interpretation. Indeed, the intercept would be equal to the predicted value of Y for $X = 0$, which does not make sense in most applications and, even when it makes sense, may be a bad case of extrapolation. For instance, in the example above, it would be the weight predicted for a height of 0 cm, a nonsense.

If we replace x by $x + 1$ in the regression equation, the predicted value changes from y to $y + b$. In practice, we interpret the slope b as the average change in Y when we switch from a case with a given X value to a case whose X value is one unit higher. In our introductory example, increasing the height 1 cm, the weight would increase, on the average, 0.94 kg.

In particular, when $b > 0$, there is a positive association between X and Y, and, when $b < 0$, a negative association. For instance, height and weight show a positive association, but, in the example of this chapter, sales and prices have a negative association (higher price results in lower sales, and conversely).

Under this perspective, the slope coefficient is sometimes called the effect of X on Y. This terminology, which is appealing, must be used with care, since the inclusion of other variables on the right side of the equation change this effect, as we will discuss in Chap. 4.

3.5 Correlation

Since a regression line never fits the data points in an exact way, it is interesting to have a numerical measure of the **goodness-of-fit**. The correlation (R) is a popular one. The interpretation is easy, because of the following properties:

- Always $-1 \leq R \leq 1$.
- The closer the absolute value of the correlation is to 1, the better is the fit (i.e. the closer the data points are to the regression line).
- The correlation and the slope of the regression line have the same sign.

How does the correlation reflect the magnitude of the residuals? A property of the regression line is that the variance of the actual values of Y is equal to the variance of the predicted values plus the variance of the residuals:

$$\text{Var(Actual values)} = \text{Var(Predicted values)} + \text{Var(Residuals)}.$$

So, simple linear regression can be seen as the decomposition of the variance of Y in these two components. As a measure of goodness-of-fit, many people prefer the squared correlation, called the **R-squared statistic**. This is due to the following property of the R-squared statistic:

$$1 - R^2 = \frac{\text{Var(Residuals)}}{\text{Var(Actual values)}}.$$

This formula allows us to take the R-squared as the proportion of the variance explained by the regression equation. Then $1 - R^2$ corresponds to the unexplained variance, that is, that of the residuals. Although the R-squared can be used instead of the correlation, it is better not to mix them, to avoid the confusion. In the examples of this book, we use always the correlation.

3.6 Obtaining the Regression Line in Excel

The intercept (a), the slope (b) and the correlation (R) are calculated in Excel with the formulas `INTERCEPT`, `SLOPE` and `CORREL`, respectively, as explained in the example that follows. Alternatively, one can start with a scatter plot, and ask the graph editor of Excel to superimpose a **linear trend**. This trend is the regression line. Under request, Excel can also display the equation and the R-squared value. The regression equation can also be obtained in Excel through the `Analysis ToolPak`, but we leave this for Chap. 4.

3.7 Example: Predicting Sales from Price

3.7.1 Presentation

Greenchips is a brand of snacks. Greenchips snacks are made of dehydrated fruits or vegetables. They are packaged in 100 g bags, as if they were potato chips, but advertised as a much healthier option. Greenchips produces snacks of dehydrated apple, pineapple and strawberry as well as chips made out of green peas or chickpeas. Their products are vegan, gluten free, with no palm oil, made of natural ingredients and oven baked instead of fried.

This example uses a sales and price data set of the Greenchips dehydrated apple snack. The data consist of weekly unit sales (thousands) of the standard 100 g package and the weekly average price (in euros) over a period of 104 weeks, starting from September 19, 2016, to September 10, 2018. The sales vary from 47 to 455 thousands of packages and the prices from 0.87 to 1.99 euros.

We develop a simple model, based on a linear equation, to predict the sales from the price. The detail of the analysis presented here can be found in the Excel file `greenchips.xls`. The data are in the sheet `Data`, with the sales in the range `B2:B105` and the prices in the range `C2:C105`. The names of the variables are in the first row.

3.7.2 The Data

Table 3.2 presents some summary statistics, obtained with the formulas `AVERAGE(B2:B105)`, `AVERAGE(C2:C105)`, `STDEV(B2:B105)`, `STDEV(C2:C105)` and `CORREL(B2:B105, C2:C105)`, respectively.

The correlation is negative and quite strong ($R = -0.881$). The negative sign is not a surprise, since the sales are expected to decrease when the price goes up. The strength of the correlation indicates that we are in a market which is highly sensitive to the price.

3.7.3 Regression Line

We take PRICE as the independent variable (X) and SALES as the dependent variable (Y). The coefficients are obtained in Excel as

```
INTERCEPT(B2:B105, C2:C105) = 626.6,
SLOPE(B2:B105, C2:C105) = -305.6.
```

Note the order of the arguments in these functions: first Y and then X. Thus, the regression equation is

$$\text{SALES} = 626.6 - 305.6\ \text{PRICE}.$$

We can see in Fig. 3.2 a scatter plot of these data, with the regression line superimposed. To obtain the scatter plot:

Table 3.2 Summary statistics ($N = 104$)

	Mean	Std. deviation	Correlation
SALES	134.66	69.50	−0.881
PRICE	1.61	0.20	

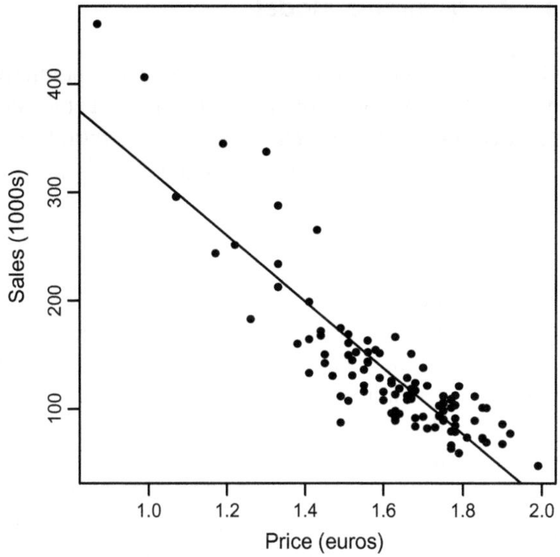

Fig. 3.2 Regression line ($R = -0.881$)

- We reorder the columns in the sheet, putting the price on the left and the sales on the right. Note that Excel takes the column on the left as the X variable and the column on the right as the Y variable.
- We select the range of the sheet containing the data.
- In the `Insert` tab, we click on the `Scatter` symbol within the `Chart` group.
- By clicking `Add Chart Element` >> `Trendline` >> `Linear`, we add the regression line to the scatter plot.
- The line can be reformatted, by selecting it, (right) mouse clicking, and then asking for `Format Trendline`. An interesting option is `Display Equation on chart`.

In Fig. 3.2, the line has negative slope (i.e. it is decreasing). Note that the sign of the slope is the same as that of the correlation, although the absolute values are quite different. Indeed, the correlation is a scale-free number, while the slope is given in thousands of packages per euro.

What is the interpretation of the regression coefficients? The intercept would be the sales corresponding to zero price, with no practical meaning. The slope may be interpreted as follows: decreasing 1% the price of the Greenchips standard package leads, on average, to an increase in the weekly sales of 3,056 packages. Note that, to avoid extrapolation, we must restrict ourselves to increments within the range of the actual data. An increment of 1% makes perfect sense here, but an increment of 1 euro is not covered by the actual data.

3.7.4 Predictive Model

Our regression equation can be regarded as a predictive model. Entering a price on the right side of our regression equation, we get a value for the sales, which we call **predicted sales**, to differentiate it from the **actual sales** that we have in our data set. The difference is the residual (note the order in the difference):

$$\text{Residual} = \text{Actual sales} - \text{Predicted sales},$$

which is taken as the error of the prediction. This is illustrated by Table 3.3. The complete results are in the sheet `Predictions`.

In Fig. 3.2, the positive residuals correspond to the points above the line, and the negative residuals to the points below. One of the properties of the regression line is that the sum of the residuals is equal to zero, so the positive and the negative residuals cancel out. Also, we see in Fig. 3.2 that the predictions are better, in general, for higher prices and lower sales. This is a situation frequently found in practice: the errors are bigger when the magnitude predicted (in this case, the sales) is higher.

How is the correlation related to the magnitude of the residuals? Since $R = -0.881$, the percentage of variance explained by the regression equation is $R^2 = 77.6\%$. Then, the percentage not explained (corresponding to the residual variance) is the complement 22.4%. This is what we get by dividing the variance of the residuals by that of the actual sales,

$$\frac{\text{Var(Residuals)}}{\text{Var(Actual values)}} = \frac{1{,}081.35}{4{,}830.80} = 0.224.$$

Table 3.3 Prediction error

Week	Price	Actual sales	Pred. sales	Residual	Percent. error (%)
1	1.60	107.804	137.713	−29.909	−27.7
2	1.67	112.659	116.323	−3.664	−3.3
3	1.51	149.850	165.213	−15.363	−10.3
⋮	⋮	⋮	⋮	⋮	⋮
103	1.75	111.933	91.878	20.055	17.9
104	1.63	98.116	128.546	−30.430	−31.0

3.8 Useful Tips

- That two variables present similar behavior (i.e. high correlation) does not imply a cause-effect relationship. Since weddings usually occur during the summer period, the correlation between temperature and number of weddings is positive. This does not mean that weddings are the cause of summer, neither that high temperatures provoke weddings. One is not the cause of the other, they simply happen together. Hence, correlation does not imply causality. *Be wary of causal statements based on correlation arguments.*
- The correlation measures the association between two variables, without considering which variable impacts the other; therefore, there is no directionality. However, in regression analysis there is directionality since we want to see, as in the example, how the independent variable (price) affects the dependent variable (sales). *Use your experience to identify which variable works better as a dependent variable.*
- The regression equation can be used for making predictions. But *extrapolation, which is making predictions out of range, should be always avoided.*

4 Multiple Regression

This chapter and Chaps. 5 to 7 cover multiple regression analysis. The discussion is oriented to practical business applications. Most technicalities have been skipped, and the mathematics have been simplified to the indispensable. So, if you have already followed a Statistics course, you may find that the attention paid to the different aspects of the regression analysis differs from what you found there.

This chapter is a mere introduction which explains how to perform a multiple regression analysis in Excel and to interpret part of the regression output. In Chap. 3, we have used an equation $Y = a + bX$ as a predictive model. Now, we include additional terms on the right side of the equation, to get a better model. Although the graphical representation of the model as a regression line is no longer possible, enlarging the equation increases the predictive power of the model and offers new insights on the data. This is illustrated by the example of this chapter, where we obtain an equation for the prediction of the resistance of a concrete from the ingredients.

4.1 Multiple Linear Regression

Loosely speaking, **linear regression** is the process by which, given data on k independent variables $X_1, X_2, \ldots, X_k$ and a dependent variable Y, we obtain the equation

$$Y = a + b_1X_1 + b_2X_2 + \cdots + b_kX_k$$

that fits the data in the best possible way. This equation is the **linear regression equation** for these data. When there is more than one independent variable, we have **multiple linear regression**. With a single independent variable ($k = 1$), we have the **simple linear regression** of Chap. 3.

A linear equation is not the only possibility for a model that predicts Y from the X variables. Methods for nonlinear regression analysis can be found in textbooks

M. Á. Canela et al., *Quantitative Methods for Management*,
https://doi.org/10.1007/978-3-030-17554-2_4

and statistical software. But this book covers only predictive models based on linear equations. So, here, regression means linear regression.

Mind that, in this context, the terms "dependent" and "independent" indicate the side of the equation on which a variable is placed. It is not assumed that the X variables are independent (in the colloquial sense) among them. On the contrary, we may find a strong association between a pair of independent variables.

4.2 Predicted Values and Residuals

A value of Y which has been calculated with a regression equation is called a **predicted value**. The difference of the actual value minus the predicted value is called **residual**:

$$\text{Residual} = \text{Actual value} - \text{Predicted value.}$$

The residuals are taken as prediction errors. So, the smaller the residuals, the better the model. In practice, the **regression algorithm** chooses the coefficients so that the sum of the squares of the residuals is minimum. This is called the **method of least squares**.

With more than one variable on the right side of the equation, implementing the least squares method involves a set of formulas which are usually packed into a matrix formula. The mathematics are hard to understand without some familiarity with matrix algebra. Assuming that you will manage this in the computer, which is quite easy to do, we omit the mathematical detail, focusing on how to perform a regression analysis in Excel, and on the output that you are going to obtain.

4.3 Interpreting the Regression Coefficients

The coefficient of a particular X variable in a regression equation is frequently interpreted as its **effect** on Y. This is based on the fact that, increasing one unit the value of that particular variable, while *holding constant the other variables*, the change in the predicted value of Y coincides with the regression coefficient. Nevertheless, we have to be careful here, keeping in mind that:

- This interpretation is bounded to situations in which it makes sense to increase or decrease a particular X variable while holding constant the rest. This is not realistic when this variable is directly related to some other variables (e.g. when the equation includes both X and the square of X), or when there is strong correlation among the independent variables.
- The coefficient of an X variable changes when another variable is added to or dropped from the equation. Sometimes this change is relevant, but sometimes it is not, as experience shows. The example of this note illustrates this point.

4.4 Multiple Correlation

As explained in Chap. 3, the regression equation never fits perfectly the data. Enlarging the equation always improves the predictions but, with real data, prediction is never exact. We use the correlation (R) between the actual and the predicted values of Y as an assessment of the **goodness-of-fit**. This correlation, which is always positive, is regarded as a numerical measurement of the association between Y and "all the X variables together". It is called **multiple correlation**, to differentiate it from the individual correlation between Y and a particular X.

The multiple R will always be positive. In a simple linear regression, it coincides with the absolute value of the correlation. Interpreting this statistic is easy: when R is close to one, the predictions given by the regression equation are good and, when it is close to zero, the predictions are poor. The intuition on how accurate the predictions are, given a correlation value, comes with practice.

Some people prefer to use R-squared instead of R. As in simple linear regression, R-squared can be interpreted as the proportion of the variance explained by the regression equation. In this book, we use always R, not R-squared, to evaluate the predictive power of a regression equation. You may decide otherwise in your practice, but, as a general rule, you should not mix both statistics, which could be confusing. As an illustration, note that, to the non-trained eye, $R = 0.5$ may seem stronger than $R^2 = 0.25$, though they are exactly the same.

In practical data analysis, we explore equations with different sets of independent variables, using the multiple correlation to compare them. In that case, we must take into account that, when an independent variable is added to a regression equation, R always increases, because the predictions are improved, although the improvement may not be relevant.

4.5 Obtaining a Regression Equation in Excel

In Excel, the sequence `Data` » `Data Analysis` » `Regression` opens the `Regression` dialog box, where we specify the variables on both sides of the regression equation. Excel allows a maximum of 16 variables on the right side of the equation. These variables have to be in consecutive columns, so that they can be specified as a block in the box `Input X Range`. If we include in the blocks specified in the boxes `Input Y Range` and `Input X Range` a first row with the names of the variables and tick in the box `Labels`, the report will include those names in the table of coefficients. If not, Excel provides names such as `X Variable 1`, `X Variable 2`, etc.

In the computer, the results of a regression analysis are presented in a report, whose format is almost universal. Excel's regression output is an example of that format. These reports are full of redundancies (e.g. Excel reports both R and R-squared). So, it is useless to look at all the figures included. In the example of this chapter, the discussion is restricted to the regression coefficients and the multiple R, since we go step by step, in order to facilitate a better understanding of the many parts of the report.

Excel's regression report contains three parts:

- In the first part (`Regression Statistics`), the main result is the multiple R.
- The second part (`ANOVA`) is a report of a statistical analysis called **analysis of variance**, which is irrelevant in business applications.
- The third part is a table of coefficients. Besides the coefficients themselves, we find the **95% confidence limits** and a **p-value** for each coefficient, which will be discussed in Chap. 5, and some intermediate calculations, like the standard error and the t statistic, with no interest in business applications.

In Table 4.1, we find the first part (complete) and the third part (a bit simplified) of the output of the second regression analysis performed in the example. Numbers have been rounded, to improve the readability, which is always recommended if you wish to share these reports with other people.

You will find that the software application that you use to obtain the regression equation (either Excel or your own choice) has an option for getting an **equation without constant**. This means that the search performed by the regression algorithm is restricted to equations with $a = 0$. This approach produces, in most cases, not only an equation with a lower predictive performance, but one for which the interpretation of some results, such as the multiple R, is more difficult. So, we assume here that you are not using this option. For a regression equation with constant term, some of the residuals are positive and some are negative, and the sum of the residuals is null.

Table 4.1 Example of regression report in Excel

Regression statistics				
Multiple R	0.789			
R square	0.623			
Adjusted R square	0.621			
Standard error	37.6			
Observations	804			
	Coefficients	P-value	Lower 95%	Upper 95%
Intercept	15.83	0.286	−13.30	44.96
Cement	0.846	0.000	0.737	0.955
Additives	0.015	0.004	0.005	0.025
Water	−0.072	0.266	−0.199	0.055

Some technicalities worth being mentioned are:

- We get an error message when there is at least one cell in the range selected whose content is not numeric (i.e. it is empty or contains text).
- If one X column is a linear function of the other X columns (this is called perfect collinearity), it is discarded, so the corresponding row in the table of coefficients is filled with zeros.

4.6 Example: Concrete Quality Control

4.6.1 Presentation

Clarice Pereira is a production manager at Cimento Costa, one of the top concrete producers in Brazil. She is in charge of supervising the quality of the production of concrete from several plants. Concrete is a main resource for construction, since it is easy to work with and is highly resistant for a variety of applications. To produce concrete, it is required to use a mixture of cement, additives, water and other components such as rock, gravel and sand.

One of the most basic ways to perform a quality control of concrete is to test its resistance 28 days after it was produced. To do this, a test is made to a concrete cylinder in which pressure is applied until the concrete breaks. An important element is to ensure the consistent quality of the concrete while taking care of the amount of cement that was used for each batch, since cement can easily represent more than 55% of the total cost of concrete.

4.6.2 The Data

Clarice plans to analyze how the ingredients of that mix impact the resistance of the concrete after 28 days. To this purpose, she gathers data on 804 production samples. The data set used for the analysis (file `concrete.xls`, sheet `Data`) contains the following variables:

- Resistance (kg/cm^2).
- Cement (kg/m^3).
- Additives (kg/m^3).
- Water (kg/m^3).

4.6.3 Regression Line (1)

We start by fitting a regression line (Resistance on Cement) to the data. The equation obtained is

$$\text{Resistance} = 0.74 + 0.99\ \text{Cement}.$$

The correlation is $R = 0.786$, which supports the idea that cement is the essential ingredient to get a high resistance. A direct interpretation of the slope coefficient suggests that, on the average, the resistance is higher by 0.99 kg/cm^2 for a concrete using 1 kg/m^3 more of cement.

Figure 4.1 is the corresponding scatter plot, with the regression line superimposed. There is a small group of points on the right side, showing that there are gaps in the range of the cement content. This will also occur for the other independent variables.

4.6.4 Regression Line (2)

A similar analysis is performed with Additives as the independent variable (Fig. 4.2). Now, the equation is

$$\text{Resistance} = 97.83 + 64.29\ \text{Additives}.$$

The correlation is $R = 0.646$, which shows that the impact of the additives on the resistance is also strong. You may wonder to what extent including the additives in the equation may improve the predictive power of the regression equation of Fig. 4.1, which is already strong.

Fig. 4.1 Resistance versus cement ($R = 0.786$)

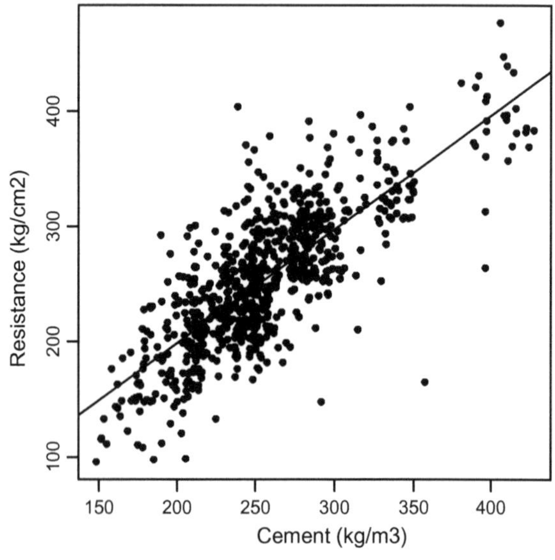

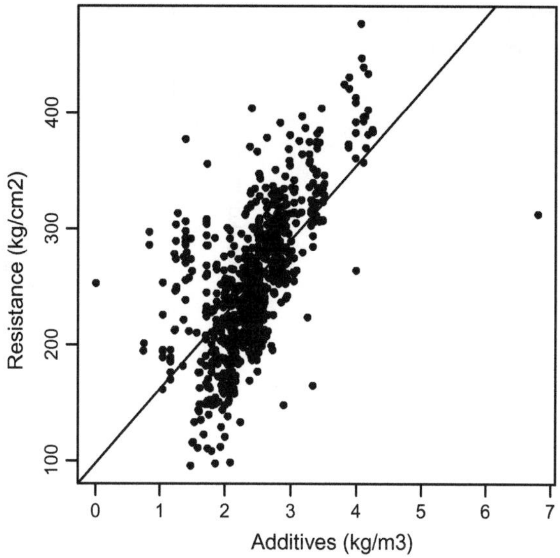

Fig. 4.2 Resistance versus additives ($R = 0.646$)

The inclusion of the point on the right of Fig. 4.2 is questionable, and it is unclear how it may affect the results obtained. So, more samples with high content in additives should be collected, or this point should be dropped from the analysis.

4.6.5 Regression Line (3)

Our third regression line is shown in Fig. 4.3. We also see here two groups of points (not the same as in Fig. 4.1). These groups point out that the water content does not vary in a continuous way across samples. This should be investigated. A potential explanation is that the data set includes data on two different products. The correlation is $R = 0.104$, showing that the impact of the water on the resistance is much weaker.

Now, the equation is

$$\text{Resistance} = 211.88 + 0.196\ \text{Water}.$$

4.6.6 Multiple Regression Analysis

In a multiple regression equation, we can use all the potential independent variables together, increasing the predictive power. The equation obtained is

$$\text{Resistance} = 15.83 + 0.85\ \text{Cement} + 15.12\ \text{Additives} - 0.07\ \text{Water}.$$

The report, as given by the Analysis ToolPak, has already been partially displayed in Table 4.1. It can also be found in the sheet `Multiple regression`.

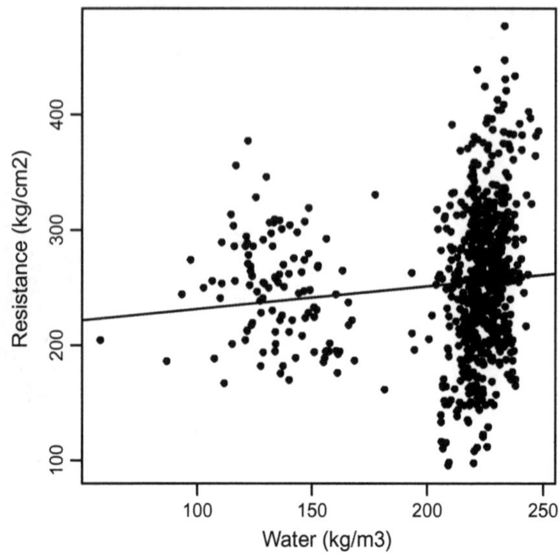

Fig. 4.3 Resistance versus water ($R = 0.104$)

Ticking in the box `Residuals` of the `Regression` dialog box, we get in the report also the predicted values and the residuals (in the range `B27:C830`). The R-squared value reported by Excel, 0.623, is the proportion of the variance of Resistance explained by the regression equation.

The multiple R reported by Excel is the square root of this number. It is equal to the correlation $R = 0.789$ of the actual and the predicted values of Resistance, illustrated by the scatter plot of Fig. 4.4. The improvement in the predictive power

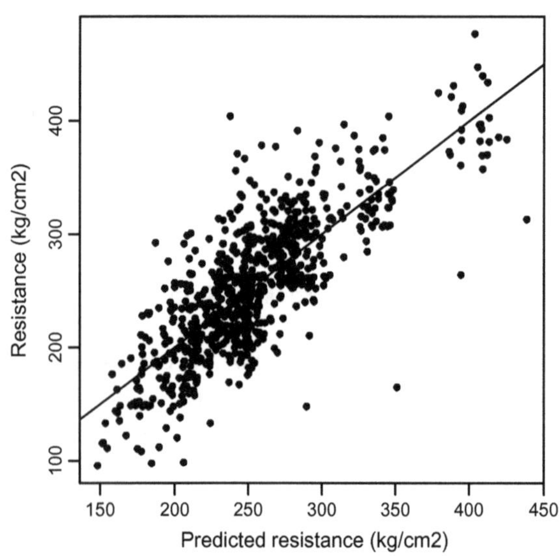

Fig. 4.4 Resistance versus predicted resistance

of the regression equation, from the first analysis to the final one, is reflected on the increase in the multiple correlation, which, in this case, is very small. Therefore, we infer that the concrete's strength is largely dependent on the amount of cement it has.

4.7 Useful Tips

- When developing a regression model, there is always a temptation to add more independent variables. *Keep that in mind that there can be a trade-off between simplicity and accuracy, and strive for the right balance.*
- Reality is complex. Specially in managerial settings, where it is almost impossible to single out a unique source of a problem. *Do not look for the perfect model, but for one that helps to take better decisions.*

Testing Regression Coefficients

5

This chapter continues the discussion of regression analysis engaged in Chaps. 3 and 4. We have seen that the regression coefficients are reported with extra information, which we have not discussed yet. How can we use this information? In particular, we explain here how to use the confidence limits and the p-values to decide about statistical significance. The example of this chapter, which deals with the analysis of the impact of the prices on market share, illustrates the interpretation of the p-values.

5.1 Statistical Inference

This chapter is mainly about confidence limits or the p-values, which are tools of **statistical inference**. The use of these tools makes full sense in a context sample/population. So, before explaining the tools, we discuss briefly the context.

Populations are usually too big to be observed completely. What we do, in practice, is to draw a **sample** from the population, perform our statistical analysis on the sample, and extend the results to the population. This is called statistical inference. Serious inference requires an assessment of the difference between a population statistical parameter and an estimate based on the analysis of a sample.

For instance, if in a survey on voting intention in the presidential elections in the US, a 42.7% of the interviewees declare their intention to vote for the Republican candidate, what can we say about the population? We take this 42.7% as an estimate of the population proportion, but acknowledging that there is an error in this figure, and that a different sample would give us a different estimate. So, the question is: how big could the error be? Confidence limits and p-values provide two ways to address this question.

M. Á. Canela et al., *Quantitative Methods for Management*,
https://doi.org/10.1007/978-3-030-17554-2_5

Without getting too technical, let us make some general remarks about statistical inference:

- First, the reliability of an estimate based on a sample depends on the **sample size**. The bigger the sample, the more we trust the results derived from that sample. In practical terms, this means that, in general, with bigger samples, we get narrower confidence intervals and lower p-values.
- For the inference from sample to population to be valid, the sample has to be extracted by means of a **random sampling** procedure. Strictly speaking, this is only achieved when the sampler has a complete list of the population units, and the selection of the sample units from that list is based on computer-generated random numbers. But, in practice, sampling is accepted as random when no part of the population is favored.
- When the sampling procedure favors a certain subpopulation, it is said to be **biased**. For instance, in an electoral survey, a sample extracted from a social network like Twitter would probably be taken as biased (toward young people), because old electors would be less represented there than in the population.

5.2 Confidence Limits

Let us suppose a population on whose individuals we can observe two variables X and Y, for which there is a linear equation $Y = a + bX$ that gives the average value of Y for a given value of X. We call this equation the **population equation**. Unless we measure the values of X and Y on every individual of the population, we cannot know exactly the population coefficients a and b.

We extract a random sample from that population, measuring X and Y on the individuals of the sample. Then, we fit a regression line to these data, using Excel or any other statistical software application. The equation of this line, derived from the sample data, will not be the same as the true population equation. We call it the **sample equation**, assuming that other samples would lead to different sample equations.

We take a sample coefficient as an estimate of the corresponding population coefficient. This estimate has an error, which can be put in evidence by collecting data from other samples and comparing, across samples, the different coefficients obtained. Nevertheless, we never do this in real-world statistical analysis. Instead, we guess how the distribution of those estimates would be.

It can be shown that, when the sample size is big enough (say $n > 100$), the sample coefficients are approximately normally distributed, with a mean which is equal to the population coefficient and a standard deviation equal to the **standard error** reported in the coefficients table of the regression output.

In practice, we take advantage of this fact by using the formula

$$\text{Coefficient} \pm (2 \times \text{Standard Error})$$

to calculate a lower and an upper limit for the regression coefficient. With a 95% confidence, we expect the true coefficient to fall between these limits. These are the **95% confidence limits** for this particular regression coefficient. The quantity after the plus/minus sign in the above formula is a bound for the error that could be made by taking the coefficient derived from the actual data as the true population value. This bound would be exceeded just in 5% of the potential samples.

Note. The factor used by your software application (Excel in our example) in the formula would not be exactly equal to 2, although very close. These factors are extracted from a probability distribution called the Student t. We skip this technicality here.

5.3 Significance

Roughly speaking, a numerical result derived from data analysis is **statistically significant** when it allows us to draw the intended conclusion. When applied to a regression coefficient derived from a sample (randomly) extracted from a population, the term significant means that we can conclude, with a certain **confidence level**, that, in the population equation, the coefficient is nonzero. There is a universal consensus about using 95% as the confidence level. So, we consider that a coefficient is significant when the corresponding 95% confidence interval does not contain zero.

Statistical significance is a key issue when the objective of the analysis is to *test a particular coefficient*, meaning to clarify whether that coefficient could be equaled to zero. This may be relevant because a coefficient being nonzero is usually taken as evidence of an influence of the corresponding independent variable on the dependent variable.

The potential effect of a particular variable is typically relevant in a scientific paper, whose conclusions are accepted only if they are based on significant results. But it may be not so in a business application whose only objective is to develop a model that produces good predictions. In this last case, we would focus on a measure of the predictive power like the multiple correlation. Also, in some applications, the analysis is not applied to prove or disprove a hypothesis about the influence of this variable on that variable, but to help in making decisions. And decisions must be made, even if the statistical analysis does not provide significance.

In general, if we are not interested in a particular independent variable, we can drop that variable from the equation when the corresponding coefficient is not significant, getting a simpler equation (in that case, we have to recalculate the whole equation).

A final comment on significance. Even disregarding the mathematics behind the p-values and the confidence limits, you should be aware that the sample size is involved in the calculations of the confidence limits. Roughly speaking, we can say that it is hard to get significance with small samples, but it is easy with hundreds of observations. You should not forget this when discussing the significance of the results. Unfortunately, significance, which is a purely statistical concept, is frequently confounded with practical relevance. The effect of a particular variable can be significant in our statistical analysis, but irrelevant in practice, when it is small.

5.4 The p-Values

Instead of the confidence limits, we can use the ***p*-value** to assess the significance of a regression coefficient. The p-value is a probability (range 0–1) which is read as follows: the lower p, the more significant the coefficient. By consensus, a coefficient is considered to be significant when $p < 0.05$ (this is equivalent to setting 95% as the confidence level).

Although p-values are used to check statistical significance, we do not seek for an interpretation of the numerical values themselves. For instance, $p = 0.315$ and $p = 0.623$ mean the same for the analyst, since both lead to the same conclusion and are too far from the threshold $p = 0.05$ to allow for doubt. Only when the p-value is close to 0.05 we may pay attention. For instance, if $p = 0.057$, we may say: "well, maybe with a bigger sample …".

Using p-values is very easy, we just check whether they are small enough. But understanding the mathematics behind them is more demanding, since calculating p-values involves integral calculus. Moreover, our intuition may not help with p-values: for instance, it is difficult to guess how dropping a specific term of a regression equation is going to affect the p-value of another term.

Note. Using $p < 0.05$ to make significance operational is based on consensus, not on this threshold having special properties. The threshold can be changed in some special applications, but, since the mathematics of p-values are not simple, we better leave that to the experts.

5.5 Multicollinearity

As we have already mentioned in Chap. 4, the regression formulas do not work when, in our data set, one of the independent variables is (exactly) equal to a linear function of the rest. Sometimes this is not exactly so, but we are very close to that situation. We say then that variable is affected by **multicollinearity**. Under multicollinearity, although the regression formulas work, the results may be unreliable.

Typically, we drop the variable affected by multicollinearity from the equation. Note that multicollinearity is frequently confounded with the fact that two independent variables are strongly correlated, which is just a particular case of multicollinearity. The practical consequences of multicollinearity are:

- Redundancy, since a variable affected by multicollinearity does not provide extra information. This means, in practice, that we may drop that variable without much loss.
- The regression coefficients can be difficult to interpret. The usual interpretation of the coefficient of a particular variable in a regression equation involves the assumption that we can change that variable while holding the other variables constant. Under multicollinearity that assumption may not make sense.
- The coefficient of the variable affected by multicollinearity can be less significant, with wider confidence intervals, making the actual estimate unreliable. We will find this in the example of this chapter. It is a technical issue, which cannot be explained without mathematical detail. But, in practice, it is good to know that this may happen, because it may help to understand the regression output when there is a multicollinearity issue.

5.6 Example: Orange Juice Pricing

5.6.1 Presentation

Minute Maid is one of the world's most famous orange juice brands. The name Minute Maid, originally created by a Boston marketing firm back in 1946, implies the convenience and ease of preparation of this orange juice. It is now produced by the Coca-Cola company, the world's leading marketer of fruit juices and drinks.

The Minute Maid sales director for Illinois is worried about Aldi, a discount supermarket that is growing fast in the Chicago area. Aldi offers its own orange juice at a very attractive price. Some managers at Minute Maid believe that Tropicana is their main competitor and, therefore, they consider that special attention should be paid to Tropicana's pricing strategy. Others believe that the real danger to Minute Maid comes from the Aldi's branded orange juice. So the sales director decides to take a closer look at the Minute Maid, Aldi, and Tropicana prices to learn how they affect Minute Maid's market share.

She will use weekly scanner data from Aldi stores in western Chicago. The current situation is as follows. Tropicana produces two kinds of orange juice, the regular juice and the premium juice, whose prices are \$3.50 and \$4.45 per package, respectively. Minute Maid and Aldi's current prices are \$3.99 and \$2.20, respectively. In this setting, Minute Maid's market share is 13.71%.

Precisely when her team is analyzing the data, the sales director receives a memo warning that Tropicana plans to reduce substantially its prices. While the price of the regular Tropicana juice would be lowered at \$3.25, two rumors circulate about the premium brand: one is that the Tropicana Premium's price is going to be set to \$3.75, and the other that it will be set at \$4.25. If Minute Maid does not respond to Tropicana's campaign, how much market share is expected to be lost?

5.6.2 The Data

The data set (file `orange.xls`, sheet `Data`) contains information on Minute Maid's market share (percentage scale) and prices (dollars per package) of Minute Maid and its competitors, covering 121 weeks. We open the analysis with a statistical summary (Table 5.1).

If the market is sensitive to price changes, Aldi looks like Minute Maid's main competitor. On average, the regular product of Tropicana is closer to Minute Maid, but a bit more expensive. The premium juice seems to be in a different market segment. Nevertheless, in the data set, the price ranges of all four brands overlap.

5.6.3 How Does Minute Maid's Price Affect Its Market Share?

We first take a simple linear regression approach to this question. The regression equation is

$$\text{MShare} = 40.01 - 6.91\ \text{MMaid},$$

with correlation $R = 0.597$. This means that an increase of 1 cent in the price leads, on average, to a loss of 0.07% in the market share of Minute Maid.

But, shan't we include the prices of the competitors, since they also could affect the market share? Table 5.2 is the table of coefficients for the corresponding regression equation. Now, $R = 0.750$, so we get additional predictive power with this enlarged equation.

Table 5.1 Summary statistics ($N = 121$)

	MShare	TropPremium	Trop	MMaid	Aldi
Mean	17.27	4.39	3.39	3.29	2.74
St. deviation	6.72	0.54	0.54	0.58	0.57

Table 5.2 Multiple regression analysis (1)

	Coefficients	Standard error	t stat	p-value
Intercept	11.41	8.29	1.38	0.172
TropPremium	8.40	6.97	1.21	0.231
Trop	−4.15	7.00	−0.59	0.554
MMaid	−8.57	0.76	−11.3	0.000
Aldi	4.13	0.74	5.58	0.000

Nevertheless, these results are not clear and somewhat counterintuitive:

- The coefficients of both Tropicana prices in Table 5.2 are nonsignificant ($p > 0.05$). Does this mean that Tropicana is not relevant here? If it were relevant, the sign of the Tropicana coefficient would be difficult to understand.
- According to Table 5.2, the impact of the Minute Maid price on the market share is stronger than the estimate derived from the equation of the regression line. Which of the two values provides a better response to our question?

5.6.4 Correlation Analysis

The first of these two points can be clarified by a correlation analysis. Table 5.3 is a **correlation matrix** for the five variables involved in our previous regression analysis. This matrix can be easily obtained through the `Analysis ToolPak`. Following the steps `Data` » `Data Analysis` » `Correlation`, we arrive at a dialog box. There, we input the range covered by the data (`B1:F122`) in the box `Input Range`. Ticking in `Labels in first row`, we get the names of the variables in the row and column names of the correlation matrix.

The strong correlation of the two Tropicana prices ($R = 0.994$) shows that we have here a clear case of multicollinearity. This could explain the nonsignificance of the coefficients of these two variables in the regression equation. Let us take a closer look at the data.

It is not hard to find in the data an explanation of that correlation. The difference between the prices of the two Tropicana juices is, most of the time, equal or close to \$1.00. This makes sense from a marketing perspective, since the customer's perception of a premium product is supported by a consistent difference in price.

5.6.5 Another Regression Analysis

The preceding discussion suggests dropping Tropicana Premium from the equation, since the coefficients of the two Tropicana juices in Table 5.2 are not only nonsignificant, but they do not have an easy interpretation. Indeed, we usually take a

Table 5.3 Correlation analysis

	MShare	TropPremium	Trop	MMaid	Aldi
MShare	1				
TropPremium	0.114	1			
Trop	0.107	0.994	1		
MMaid	−0.597	0.287	0.298	1	
Aldi	0.225	−0.048	−0.028	0.140	1

Table 5.4 Multiple regression analysis (2)

	Coefficients	Standard error	t stat	p-value
Intercept	20.42	3.60	5.67	0.000
Trop	4.23	0.81	5.23	0.000
MMaid	−8.63	0.76	−11.4	0.000
Aldi	3.98	0.73	5.45	0.000

regression coefficient as an assessment of the impact of the corresponding variable on the dependent variable *when the other independent variables are held constant.* But this setting is too unrealistic for the two Tropicana terms, according to our data.

Table 5.4 is the table of coefficients for a reduced regression equation. The multiple correlation is, now, $R = 0.746$, practically the same. All the coefficients are significant, and the signs make sense.

Once the first of the points raised by the results of Table 5.2 has been addressed, and, feeling more comfortable with the results of Table 5.4, we go for the second point. While the slope of the regression line was −6.91, the coefficient of MMaid in Table 5.4 is −8.63. Which is the right one?

Both are right, responding to different questions. The coefficient of Table 5.4 gives us the impact on the market share of a change in the price of Minute Maid, assuming that the prices of the competitors are held constant. If this assumption is dropped, the right assessment of the impact of Minute Maid prices is the coefficient of the regression line.

5.6.6 What Is the Impact of Minute Maid's Price?

If we have to respond to this question in a realistic way, we should use the regression line, because, in the real world, we do not see prices of one juice changing while the prices of the competitors remain constant. At least, we do not find that in our data. Moreover, the correlations of Table 5.3 show that the prices of Aldi and Tropicana change in the same direction of those of Minute Maid.

All right, but, then, what do we assume about Aldi and Tropicana's prices when using the regression line? Nothing, that is, that they behave as usual (in our data set). This makes the model applicable.

5.6.7 What Will Happen if Minute Maid Does Not React to Tropicana's Move?

For the first rumor, we cannot say much. We do not have data for which the difference between Tropicana and Tropicana Premium prices is as high as \$0.50. Actually, we regard Tropicana Premium as a product for another customer segment but, if the difference with the regular Tropicana juice is halved, who knows?

Table 5.5 Multiple regression analysis (3)

	Coefficients	Standard error	t stat	p-value
Intercept	30.31	3.47	8.74	0.000
Trop	3.90	0.90	4.34	0.000
MMaid	−7.99	0.83	−9.60	0.000

For the second rumor, what matters is the change in the regular Tropicana's price. The coefficient of Tropicana in Table 5.4 would be valid for addressing this question if we could also assume that Aldi does not react to Tropicana's move. Then, for a price cut of $0.25, we would get a drop in the market share of

$$0.25 \times 4.230\% = 1.06\%.$$

We may think that assuming that Aldi will not react to Tropicana's move is not realistic (given the correlations of Table 5.3, it is quite unrealistic). Then, the equation of Table 5.4 would not be adequate, since we need one in which Aldi is not included.

Dropping Aldi from the equation, we get the results of Table 5.5, with multiple correlation $R = 0.667$. Now, our estimate of the drop in the market share would be

$$0.25 \times 3.905\% = 0.97\%.$$

5.7 Useful Tips

- Working with a sample allows us to make inferences about the population without having to make the expenditure, both in time and in resources, in order to study the full population. Generally speaking, the bigger the sample, the better, but also the more expensive. Remember that *precision has a cost.*
- *Avoid convenience sampling when planning your quantitative analysis.* It is easy to interview customers who are easily accessible and to neglect those that are difficult to reach. With a convenience sample, we can have biased results that do not reflect the population under study.
- *Decisions must be made, even if the statistical analysis does not provide significance.* Quantitative analyses should support decision-making, but should not be a substitute for managerial interventions.
- *Clarify what you are expecting from a regression model.* Sometimes, you are interested in knowing if a variable has a negative or positive impact on another one. Then, your focus would be on the sign and the significance of the coefficient. In other cases, you are interested in the strength of the impact. Then, you would focus on the magnitude of the coefficient or in the overall accuracy of the model.
- Under multicollinearity, you may obtain results which do not look right. *Double check the analysis, since your instinct can let you know that something is wrong in the calculations.*

Dummy Variables

6

In this chapter, we explain how to introduce categorical variables in a regression analysis, coding the categories with dummy variables. This is needed in most of the applications of regression analysis, since the samples on which we collect our data are typically partitioned into groups. In the example, we use a dummy variable to code gender, which allows us to include the comparison between genders in the analysis in an easy way.

6.1 Dummy Variables

Quite often, the data that we analyze include **categorical variables**, which split the sample into groups. Examples of categorical variables are gender, marital status, citizenship, industrial sector, etc. To include the groups in the regression equation, we use **dummy variables**.

A dummy variable or, more briefly, a dummy, is a variable taking values 0 and 1. In regression analysis, dummies are used to code groups. In this chapter, we explain how to code groups with dummies, and how to interpret the coefficients of those dummies in a regression equation.

6.2 Coding Two Groups with a Dummy

Let us start with the simplest case, in which we have only two groups, for instance, female and male managers. We code these two groups with a dummy D, which takes values as follows: $D = 1$ in the male group, and $D = 0$ in the female group. We wish to interpret the coefficients of an equation

$$Y = a + b\,D,$$

M. Á. Canela et al., *Quantitative Methods for Management*,
https://doi.org/10.1007/978-3-030-17554-2_6

in which Y can be any continuous variable, such as the salary in the example of this chapter.

Since D can only take two values, the equation can be easily read by looking at the two cases separately:

- In the group $D = 0$ (female), the predicted value of Y is a.
- In the group $D = 1$ (male), the predicted value of Y is $a + b$.

So, the intercept a is the average value of Y in the female group, and the slope b is the average difference between the two groups. If we wish to test the average difference between the two groups, we can do it by testing the slope coefficient. This idea is used in the example for testing the existence of a gender gap in wages.

The interpretation is a bit different when there is an additional variable X in the equation,

$$Y = a + b\,D + c\,X.$$

Now, according to what we have seen in Chaps. 4 and 5 concerning the interpretation of the coefficients in a multiple regression equation, the coefficient b is interpreted as the average difference between the two groups *for a given value of* X.

Since we have coded male as 1 and female as 0, the difference is read as male minus female. Coding the genders in the opposite way, the slope would have the opposite sign and would be read as female minus male. It is usually good practice to name a dummy variable as the group where it takes value 1. So, the dummy of this example could be called MALE. This is, in general, better than calling it GENDER, which needs additional clarification (see the example).

6.3 Three Groups

If there are more than two groups, we need more than one dummy to code the groups. More specifically, we need as many dummies as the number of groups minus one: we set one of the groups as the baseline group, and create one dummy for each of the other groups.

We show this with a three-group example (Table 6.1). Imagine that a sample of managers is partitioned, based on marital status, into three groups: single, married, and divorced. We take (arbitrarily) the singles as the baseline group, creating two dummies, D_1 and D_2, as follows: (a) $D_1 = 1$ in the married group and $D_1 = 0$ in the rest, and (b) $D_2 = 1$ in the divorced group and $D_2 = 0$ in the rest.

Thus, the three groups get coded as follows:

- The single group corresponds to $D_1 = D_2 = 0$.
- The married group corresponds to $D_1 = 1$ and $D_2 = 0$.
- The divorced group corresponds to $D_1 = 0$ and $D_2 = 1$.

Table 6.1 Coding three groups

Group	D_1	D_2
Single	0	0
Married	1	0
Divorced	0	1

Now, let us see how to interpret the coefficients of a regression equation

$$Y = a + b_1\, D_1 + b_2\, D_2.$$

As with two groups, we look at each group separately. The average values of Y on the three groups are, respectively: a (singles), $a + b_1$ (married) and $a + b_2$ (divorced). So, the coefficient b_1 would be the average difference married minus single, while b_2 would be the average difference divorced minus single.

6.4 Clarification

Although it looks natural to code gender with a single variable, using two dummies for three groups may look counterintuitive. The following two arguments can help to understand how this works.

- As illustrated by the above example, three groups are unequivocally identified with two dummies.
- Suppose that we create an additional dummy D_0 for the zero group (singles in Table 6.1) and we ask the computer to fit an equation

  $$Y = a + b_0\, D_0 + b_1\, D_1 + b_2\, D_2$$

  to the data. Since we have $D_0 + D_1 + D_2 = 1$ for all the managers in the sample, the three dummies are not linearly independent, so the regression algorithm will not admit the three of them together.

6.5 Any Number of Groups

The preceding discussion is easily extended to an arbitrary number of groups $k > 1$. We select one group as the baseline, creating the $k - 1$ dummies associated with the other groups. All the groups get coded, as shown in Table 6.1, and the coefficients of the dummies in a regression equation are interpreted as mean differences between groups.

Of course, the interpretation of the coefficients depends on the choice of the baseline. In practice, our choice would be aligned with our analysis. In the example, we are concerned with the gender salary gap, which is typically referred to as a difference male minus female, so we take the female group as the baseline.

6.6 Example: Gender Salary Gap

6.6.1 Presentation

Scandia, a big Norwegian corporation with an annual turnover of $5,000 million and nearly 30,000 employees worldwide, has a long history of respect for diversity. Back in 1976, Scandia was the worldwide pioneer in establishing an internal rule against gender, race, and religious discrimination. Since then, it has been a model for managing diversity in organizations, and Scandia's CEO has been frequently invited to explain the corporation's employee diversity policy.

In the last decade, Scandia observed that the number of women in senior management positions was very low, and decided to make an effort to, whenever possible, increase the proportion of women as board members, regional managers, country managers, and heads of department.

Preparing a meeting in Bogotá, where she will meet with some country managers of Latin America, an HR manager reviews the actual data on salaries for the Latin American managers. It seems that Scandia has done a good job in promoting women in the region: a relevant proportion of senior managers are women. But, although this is good news, it turns out that their average salary is much lower than that of men!

At the regional headquarters, she meets Colombia's country manager. Discussing what she has found, the country manager suggests that the salary gap is not due to gender. According to him, the source of the difference is that, in general, the female managers have been less time with the company, so they had fewer opportunities to get salary increases.

6.6.2 The Data

The data for the analysis (file `gender.xls`, sheet `Data (1)`) covers 288 managers. The gender distribution is, indeed, balanced, with 140 female and 148 male managers. The salaries are in US dollars, and there is also data on the tenure, in years.

6.6.3 How Wide Is the Gender Salary Gap?

In Excel, the group means can be obtained in a **pivot table**. To get this:

1. We select the range containing the relevant data (`B2:D289`).
2. In the `Insert` tab, we click the `PivotTable` button of the `Tables` group.
3. We click `PivotTable` in the drop-down menu, and Excel opens the `Create PivotTable` dialog box.
4. We drag GENDER to the `Row Labels` area and SALARY to the `Values` area.
5. We change the calculation by right-clicking `Sum of SALARY`, selecting `Field setting`, and switching to `Average`.

The average annual salary is \$74,420 for the female managers and \$91,552 for the male managers. Indeed, this looks like a clear case of gender discrimination.

6.6.4 Can the Gap Be Explained by the Number of Years with the Company?

In the pivot table, we replace SALARY by TENURE. This gives us the average tenure for the two groups: 12.6 years for the female managers and 17.5 years for the male managers. So, male managers have, on the average, been longer than female managers in the company, as remarked by the Colombia's country manager.

Now, the regression of salary on tenure (Fig. 6.1) shows that managers with more time in the company have higher salaries. The correlation between tenure and salary

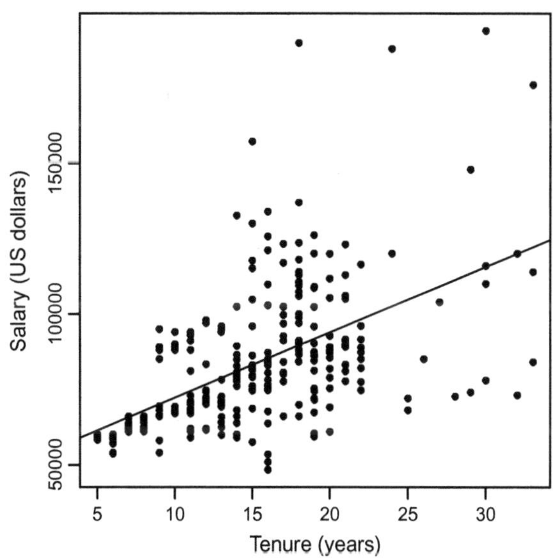

Fig. 6.1 Salary versus tenure ($R = 0.549$)

is positive and relevant, $R = 0.549$. This also supports the argument of Colombia's country manager. May it be that the difference in tenure accounts for the salary gap?

6.6.5 Regression Analysis (1)

For sure, the salary gap looks relevant to us, but is it statistically significant? In plain words, if there was not any gender gap and we assigned salaries at random, could we get such a high mean difference? We see now how to address this question through regression analysis.

As explained in the first part of this chapter, regressing SALARY on a dummy for being male, we get an equation whose slope coincides with the mean difference in SALARY between the two genders.

To perform this regression analysis, we create the gender dummy, which we call MALE, with the function `IF`. In column `D` of sheet `Data (1)`, we have the gender. Now, we introduce in `E2` the formula `IF(D2 = "Male", 1, 0)`, copying it down. In this way, we get the dummy in the range `E2:E289`. Instead of leaving it in column E, we have transferred the dummy to a new sheet. So, in the sheet `Data (2)` we have a new, clean data set.

Table 6.2 is the coefficients table for the regression of SALARY on MALE. A scatterplot would not be useful here, since a regression line does not make sense when the independent variable is a dummy. The intercept is equal to the mean salary of the female group, as obtained in the pivot table, and the slope is equal to the mean difference male minus female. The p-value of the slope coefficient is practically zero.

6.6.6 Regression Analysis (2)

The advantage of a regression approach is that it allows us to include the three variables together. Table 6.3 shows the results of the regression of SALARY on

Table 6.2 Regression analysis (1)

	Coefficients	Standard error	t stat	p-value
Intercept	74,419.86	1,802.03	41.3	0.000
MALE	17,132.11	2,513.79	6.82	0.000

Table 6.3 Regression analysis (2)

	Coefficients	Standard error	t stat	p-value
Intercept	50,725.86	3,115.84	16.28	0.000
TENURE	1,884.75	212.75	8.86	0.000
MALE	7,906.66	2461.01	3.21	0.001

TENURE and MALE. The correlation is, now, $R = 0.570$. Although it is still highly significant, the coefficient of MALE here is less than one half of that of Table 6.2.

What is the interpretation of this new coefficient? It has to be interpreted as the change in salary due to a change in the variable MALE, holding the other variables (that is, TENURE) constant. So, it tells us that, for the same number of years in the company, the female managers get, on average, \$7,907 less. So, even if TENURE explains, in part, the salary gap, there is still one part of that which could be attributed to gender.

6.7 Useful Tips

- Categorical variables are frequently involved in regression models, since they facilitate the study of the difference between groups. *Always transform categorical variables into dummies to include them in the regression equation.*
- Some numeric variables might be misleading, they look like numbers but in reality, they represent categories. Months, weekdays or SKU codes are good examples of that. January is frequently expressed as month 1 and February as month 2 but these numbers do not have any meaning as real numbers. January plus February means nothing. *Do not get confused with the numeric values that are used to code categorical variables.*
- *Continuous variables can be transformed into dummies.* This process is called binarization. For example, you might have a variable that contains information about the size of houses. You could divide the houses into "big" and "small" by simply choosing a cutoff point. The variable can then be converted into a dummy (big = 1, small = 0).
- If a categorical variable is transformed into several dummies, *always leave out of the regression equation one of the dummy variables.* The interpretation of the coefficient of a dummy will then be the mean difference between the group associated with that dummy and the group whose dummy has been left out.

Interaction 7

This chapter adds sophistication to the regression equation through interaction terms. By including an extra term in the equation, we allow for the effect of an independent variable to change with the level of another independent variable. In the example, we use an interaction term to account for the effect of some variables on the efficiency of a truck to be different for different types of motor.

7.1 Interaction in a Regression Equation

Let X and Z be two variables on the right side of a linear regression equation and Y the variable on the left side. When the effect of X on Y depends on the value of Z, we say that there is an **interaction effect** of X and Z on Y. An interaction effect can be included in a regression equation through a product term, as in

$$Y = a + bX + cZ + dXZ.$$

We identify the interaction effect with the coefficient d. In mathematical terms, the interpretation of the coefficient is easy. Writing the above equation as

$$Y = \big(a + cZ\big) + \big(b + dZ\big)X,$$

we see the effect of X on Y as $b + dZ$, so it depends on the value of Z.

Nevertheless, in practice, one has to be careful with product terms. Let us point out some practical issues:

- There is no way to interpret separately the three terms in which X and Z are involved.

M. Á. Canela et al., *Quantitative Methods for Management*,
https://doi.org/10.1007/978-3-030-17554-2_7

- The best way to understand an interaction effect is to assign to one of the two factors the role of a **moderator** of the effect of the other, as we did above. We looked at the equation so that the effect of X on Y was different for different values of Z. This role is arbitrary, since we can also say that X moderates the effect of Z on Y. In practice, moderation roles are assigned according to the orientation of the analysis.
- The interaction is easy to describe when the variable taken as the moderator is a dummy, as we see below.

7.2 Interpretation of an Interaction Term

Let us suppose that, in the equation of Sect. 7.1, Z is a dummy variable which codes two groups (e.g. regular/special). To interpret the coefficient of the product term, we take the two groups separately:

- In the group $Z = 0$, the equation becomes $Y = a + bX$, so the effect of X on Y is equal to b.
- In the group $Z = 1$, we have $Y = (a + c) + (b + d)X$, so the effect of X on Y is $b + d$.

Therefore, d accounts for the change in the effect of X across groups, that is, for the interaction effect. We apply these ideas in the example that follows.

7.3 Example: Diesel Consumption

7.3.1 Presentation

Martorell Transportes is a Mexican company transporting perishable goods in the northeast area of the country. Since the refrigerated transportation of perishables is hyper competitive, any insight in terms of cost efficiency would help Martorell to gain a competitive advantage.

Over the past years, the cost of diesel has increased substantially, so the company would like to understand better what can be done to improve the efficiency in the use of diesel fuel. This is a critical project for Martorell, since fuel accounts for about 32% of their freight costs.

Martorell's truck fleet includes a collection of older trucks, with Daimler engines together with some newer, more expensive vehicles, with other engine types. Daimler engines are thought to be less efficient.

7.3.2 The Data

The company's controller provides data on efficiency for a collection of trips made by Martorell's trucks (file `diesel.xls`, sheet `Data`). The data set contains the following variables:

- EFFICIENCY: the fuel efficiency, calculated as the ratio of the distance traveled in kilometers to the diesel consumed in liters (km/l).
- IDLE: the percentage of time the truck's engine is not moving over the total time that it is turned on.
- POWER: the average percentage of use of the engine's power.
- DAIMLER: a dummy for the truck having an engine from Daimler Trucks.
- YEARS: how old the truck is, calculated as the current year minus the year the truck's model was released.

Martorell management understand that engine's idle time and power use can be controlled by the truck's driver, but the engine type and the age of the truck model brand cannot. They also think that the use of the engine's power would be the best predictor. A regression equation predicting the fuel efficiency in terms of the other variables could be useful here.

7.3.3 Regression Line (1)

We start by fitting a regression line (EFFICIENCY on POWER) to the data. The equation obtained is

$$\text{EFFICIENCY} = 3.212 - 0.0086\ \text{POWER}.$$

Figure 7.1 is the corresponding scatter plot. The correlation is $R = -0.339$. A negative correlation means that the efficiency decreases when the trucks use more power. Nevertheless, Fig. 7.1 shows that this is not that clear.

In the scatter plot, two clusters are clearly revealed. It is easy to check that they correspond to the two values of DAIMLER, that is, to the two types of engine. The fuel efficiency seems to be a bit better on the left cluster, but it is unclear whether this is due to the engine's type or to using less power. So, it would be better to include the type in the analysis, as we will do later.

7.3.4 Regression Line (2)

A similar analysis is performed with IDLE as the independent variable. Now, the equation is

$$\text{EFFICIENCY} = 3.133 - 0.0075\ \text{IDLE}.$$

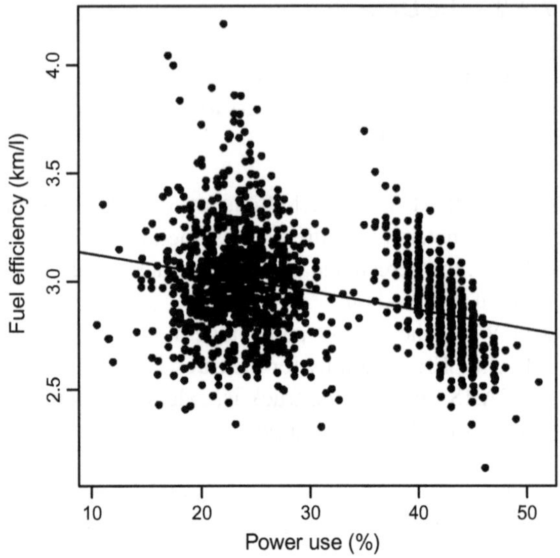

Fig. 7.1 Fuel efficiency versus power use ($R = -0.339$)

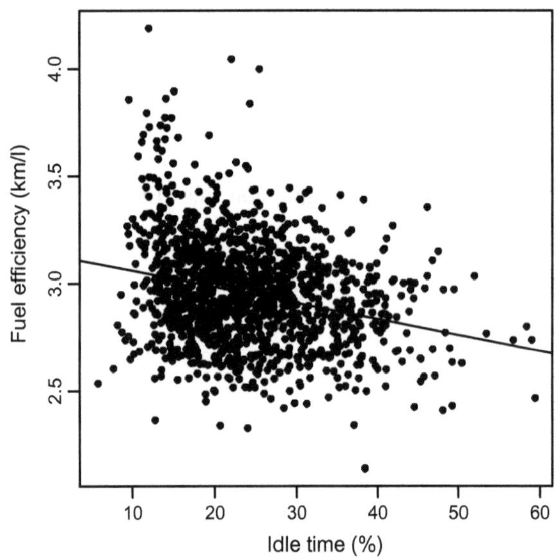

Fig. 7.2 Fuel efficiency versus idle time ($R = -0.257$)

The correlation is weaker here, and also negative, $R = -0.257$. This equation predicts, for trucks having a 1% more of idle time, an efficiency ratio decrease of 0.0075. The scatter plot is shown in Fig. 7.2.

7.3.5 Multiple Regression Analysis

In a multiple regression equation, we can use all the potential independent variables together, increasing the predictive power. Indeed, the multiple correlation is, now, $R = 0.655$.

The results (Table 7.1) can be found in the sheet `Multiple Regression` of the Excel file. Note that both idle time and power use seem to have a stronger effect now. The sign of the coefficient of DAIMLER suggests that, for the same age, trucks with a Daimler engine are less efficient.

7.3.6 Splitting the Sample by Motor Type

Figure 7.1 suggests that the effect of POWER on EFFICIENCY could be different for the two types of engine. So, it could be interesting to split the data set in two subsets and perform separate analyses. The sheets `Data Daimler` and `Data Other` contains these two parts. The respective sample sizes are 859 and 462.

We have gathered in Table 7.2 average values for the efficiency ratio, the idle time and the percentage of power utilization in the two subsamples. The average efficiency is higher for the Daimler motors, which agrees with Fig. 7.1. There is also a small difference in idle time, and a big gap in utilization, which creates the two clusters of Fig. 7.1. The values within parentheses are standard deviations.

Table 7.3 shows that the age difference between the two types of engine. The truck models with a Daimler engine are 4 or more years old, while most of other models are 2 or 3 years old.

Table 7.1 Multiple regression analysis ($N = 1{,}321$)

	Coefficients	Standard error	t stat	P-value
Intercept	5.379	0.097	55.5	0.000
IDLE	–0.020	0.001	–25.7	0.000
USE	–0.047	0.002	–23.9	0.000
DAIMLER	–0.659	0.040	–16.4	0.000
YEARS	–0.024	0.006	–3.73	0.000

Table 7.2 Average values per engine type

	Daimler	Other
EFFICIENCY	3.006 (0.255)	2.859 (0.194)
IDLE	24.88 (8.82)	22.34 (7.44)
POWER	23.28 (3.63)	42.33 (2.52)

Table 7.3 Age by motor type

	1	2	3	4	5	6
DAIMLER				361	398	100
OTHER	8	207	145	52	50	

Table 7.4 Regression analysis for Daimler engines ($N = 859$)

	Coefficients	Standard error	t stat	P-value
Intercept	5.657	0.098	58.0	0.000
IDLE	–0.031	0.001	–27.6	0.000
POWER	–0.065	0.003	–23.8	0.000
YEARS	–0.076	0.009	–8.17	0.000

Table 7.5 Regression analysis for other engines ($N = 462$)

	Coefficients	Standard error	t stat	P-value
Intercept	5.366	0.108	49.9	0.000
IDLE	–0.007	0.001	–8.32	0.000
POWER	–0.055	0.002	–22.8	0.000
YEARS	–0.005	0.006	–0.90	0.371

The figures in these two tables support the separate analyses in the two subsamples. Table 7.4 presents the regression analysis for Daimler motors. The correlation is a bit stronger than in Table 7.1, $R = 0.706$. All the coefficients are significant.

In Table 7.5, we have the results of the regression analysis for the other engines. The correlation is stronger than in Table 7.4, $R = 0.742$. Not all the coefficients are also significant here.

Comparing Tables 7.4 and 7.5, we find that:

- The effect of IDLE is four times stronger (with the same sign) in the Daimler motors.
- The effect of POWER is of the same order of magnitude.
- The effect of YEARS is much stronger in trucks with a Daimler engine. This could be due, in part, to the fact that they are older.

7.3.7 Analysis with Interaction Terms

The inclusion of product terms allows us to combine the two separate analyses of Tables 7.4 and 7.5 in a single analysis, testing the significance of the differences between the two tables that we have mentioned above. Table 7.6 presents the results

Table 7.6 Regression analysis with interaction terms

	Coefficients	Standard error	*t* stat	*P*-value
Intercept	5.366	0.136	39.4	0.000
DAIMLER	0.290	0.163	1.78	0.075
IDLE	–0.007	0.001	–6.56	0.000
DAIMLER * IDLE	–0.024	0.001	–16.3	0.000
POWER	–0.055	0.003	–18.0	0.000
DAIMLER * POWER	–0.010	0.004	–2.55	0.011
YEARS	–0.005	0.008	–0.71	0.480
DAIMLER * YEARS	–0.070	0.011	–6.12	0.000

obtained. The correlation is $R = 0.742$. Note that, although not all the terms are significant, the product terms are.

As an illustration of how to interpret the coefficients in this type of equation, let us take the two terms involving IDLE. For the non-Daimler models, the effect is -0.007, meaning that, for trucks having a 1% more of idle time, the efficiency ratio is 0.007 less. But for Daimler, the same increase in idle time leads to a decrease in the efficiency ratio of $0.007 + 0.024 = 0.031$. The interpretation is similar for POWER and YEARS.

7.4 Useful Tips

- *Interaction and moderation refer to the same phenomenon*, which is that the effect of one variable depends on the value of another variable.
- In the example of this chapter, the interaction term was the product of a dummy variable and a continuous variable. Interaction terms can also be the product of two dummies or the product of two continuous variables. In the latter case, *the interpretation is more complex*.
- When including in a regression equation an interaction term involving two variables, *remember that you cannot interpret separately the three coefficients related to these variables*.

Part III
Classification

Classification Models

8

This chapter deals with classification models. As in regression analysis, the objective is to predict a dependent variable (Y) from a set of independent variables (X's). The difference is that, in classification models, Y is a categorical variable. This book only covers binary classification, in which Y takes two values. The example of this chapter is a typical application, default prediction.

8.1 Classification Models

In a **classification** setting, we have a population divided into several **classes** or groups, and we wish to predict group membership (Y) from a set of X variables. Classification models help us to deal with many complex situations in business. Some popular examples that illustrate this are:

- **Churn modeling**. The term churn is used in marketing to refer to a customer leaving the company. A key step in proactive churn management is to predict whether a customer is likely to churn, since an early detection of the potential churners helps to plan a retention campaign. For instance, a mobile telephone company may wish to classify its customers as either churners or non-churners (Y), based on demographics and consumption habits (X's).
- **Direct marketing campaign**. There are two main approaches for companies to promote products and/or services: through mass campaigns, targeting general indiscriminate public, or through direct marketing, targeting a specific set of contacts. Since positive response to mass campaigns is typically very low, direct marketing focuses on targets that assumably will be keener to that specific

M. Á. Canela et al., *Quantitative Methods for Management*,
https://doi.org/10.1007/978-3-030-17554-2_8

product/service. A classification model that predicts positive response helps the campaign manager to select the target customers.
- **Spam filters**. A spam filter is an algorithm which classifies e-mail messages as either spam or non-spam, based on a collection of attributes such as the occurrence of certain words or characters in the messages.

Churn modeling and direct marketing campaigns apply classification methods in the context of **Customer Relationship Management** (CRM). Other classics are **fraud detection** and **credit scoring**. The example of this chapter deals with a situation close to that of credit scoring, since we try to identify potential loan defaulters.

8.2 Binary Classification

In the simplest version of the classification problem, there are only two classes, which are typically coded with a dummy variable. This is called **binary classification**. In this book, we only deal with binary classification. The class corresponding to $Y = 1$ is usually called **positive** and the class $Y = 0$, **negative**. Of course, these labels are arbitrary, and can be switched to facilitate the understanding of the users of the classification model.

A binary classification model produces, for every combination of the X variables, a numeric value, called **predictive score** (also propensity score). The various classification methods differ on how they calculate the scores. In this book we use a linear equation for this purpose, taking the predicted values given by the equation as scores.

The effective classification, irrespective of the method used to produce the score, is performed by comparing the score with a cutoff value, classifying as positives those cases for which the predicted value exceeds the cutoff, and as negatives the rest of the cases:

- SCORE > CUTOFF $\Longrightarrow$ PREDICTED POSITIVE.
- SCORE < CUTOFF $\Longrightarrow$ PREDICTED NEGATIVE.

This is easily performed in Excel with the function `IF`. Although 0.5 may look as the obvious cutoff, it is frequently replaced by other choices, based on the specific context in which the model has to be applied. In a business context, it may happen that the choice of the cutoff is supported by a **cost/benefit analysis**.

Although linear regression is the simplest approach, the scores do not necessarily fall within the 0–1 range. More elegant is the solution of an alternative method, called **logistic regression**, which produces scores in that range. The mathematics involved in the obtention of the equation are beyond the scope of plain Excel. Nevertheless, logistic regression is available in statistical packages like SPSS or Stata, and in the current languages used by data scientists, R and Python.

8.3 Confusion Matrix

Irrespective of the method used to develop it, a classification model can be evaluated in a very simple way, based on a **confusion matrix**. The confusion matrix results from the cross-tabulation of the actual class in the data set and the predicted class obtained as explained in the preceding section. Although there is no universal consensus, the predicted class typically comes in the rows and the actual class in the columns.

In a binary classification context, the cross-tabulation of actual and predicted class yields four situations, called true positive (TP), false positive (FP), false negative (FN) and true negative (TN), as shown in Table 8.1.

The proportion of cases classified in the right way is called the **accuracy**:

$$\text{Accuracy} = \frac{\text{TP} + \text{TN}}{\text{TP} + \text{FP} + \text{FN} + \text{TN}}.$$

Although the accuracy can be used as a summary statistic, it is not always the main concern, since some of the terms of this matrix may have a special relevance. Other measures which can be prioritized, depending on the application, are:

- The **true positive rate**, or proportion of right classification among the actual positives,

$$\text{TP rate} = \frac{\text{TP}}{\text{TP} + \text{FN}}.$$

- The **false positive rate**, or proportion of wrong classification among the actual negatives,

$$\text{FP rate} = \frac{\text{FP}}{\text{FP} + \text{TN}}.$$

Typically, we wish the TP rate to be high, and the FP rate to be low. But in real applications, they are not regarded as equally important. For instance, in churn modeling, to have a respectable TP rate is capital, since we aim at detecting potential churners, and taking false alarms as a lesser evil. But in a spam filter, where spam messages (positives) are filtered out so only legal messages (negatives) come directly to us, the FP rate has to be very low, since we do not want legal messages to be put in a separate folder so that we have to search for them.

Table 8.1 Confusion matrix

	Actual positive	Actual negative
Predicted positive	TP	FP
Predicted negative	FN	TN

8.4 Example: Default at Alexia Bank

8.4.1 Presentation

Alexia is a big European bank, with an office very close to the European Commission headquarters in Brussels. This makes this office the one with the biggest number of international clients. Alexia has country-specific credit scoring models, which they use for accepting or rejecting loan requests. But the manager of this particular office feels that, given the specificities of his clients, he should use a different model.

His analysis is restricted to clients having a job, and is focused on nine countries: Denmark, France, Germany, Greece, Holland, Italy, Poland, Spain and United Kingdom. From a database of clients that have previously asked for a loan, he filters out those clients from other countries. Then he extracts from the resulting list of clients a random sample of size 3,000. With some selected attributes, he plans to develop a model for predicting default.

8.4.2 The Data Set

The data set can be found in the sheet `Raw data` of the Excel file `alexia.xls`. The variables are:

- DEFAULT (Yes/No), whether the client defaulted.
- GENDER (Male/Female).
- AGE, the age in years.
- CITIZEN (DE/FR/GE/GR/HO/IT/PO/SP/UK), the client's citizenship.
- PART (Yes/No), whether the client had a part-time job.
- INCOME, the client's annual income, in euros.
- BALANCE, the client's balance when the application for the loan was filed, in euros.

8.4.3 Data Description

We start the analysis with two 2×2 tables, which can be easily produced as pivot tables in Excel. Table 8.2 shows that, indeed, gender can help to predict default, since the default rate is higher among the male clients (12.8%) than among the female clients (6.4%).

Table 8.2 Default by gender

	Default	Non-default
Male	282	1,925
Female	51	742

Table 8.3 Default by work schedule

	Default	Non-default
Part time	127	770
Full-time	206	1,897

In a similar way, we find a higher default rate among clients working part-time (14.2%) than among clients working full-time (9.8%). But, even if these findings are interesting, we are not satisfied with these 2 × 2 tables, because we think that a model using all the information available can produce better predictions (Table 8.3).

8.4.4 Regression Analysis

Our classification model is based on a linear regression equation, with DEFAULT on the left side. To carry out the analysis in Excel, DEFAULT has to be transformed into a dummy. We use the formula `IF(A2 = "Yes", 1, 0)` for that. The same for PART, with the formula `IF(E2 = "Yes", 1, 0)`. In the same way, we replace GENDER by a dummy created with `IF(B2 = "Female", 1, 0)`.

Finally, we have to create a set of dummies to deal with CITIZEN. As we saw in Chap. 6, we have to include in the equation as many dummies as the number of groups minus one (8). We take DE as the baseline group (this choice is irrelevant for the actual analysis). For FR, we use `IF(D2 = "FR", 1, 0)`. The same for GE, GR, HO, IT, PO, SP, and UK.

This leaves us with 13 independent variables, which, as discussed in Chapter 4, can be managed in Excel by the `Analysis ToolPak`. The new data set can be found in the sheet `Transformed data`. Table 8.4 is the coefficients table for the linear regression of DEFAULT on the rest of the variables of the new sheet.

The multiple correlation is $R = 0.617$. Since the focus is put on classification, we do not use the correlation to assess how effective our model is, but a confusion matrix. Also, we do not discuss the significance of the regression coefficients.

8.4.5 Classification

We classify the clients as defaulters or non-defaulters, using 0.5 as a cutoff. To do this, we include the predicted values in the regression output, by marking the tick box `Residuals` in the dialog window of the function `Regression` of the `Analysis ToolPak`. The results, a bit edited, have been collected in the sheet `Regression results`. The values of `Predicted Y` are in the range `B29:B3028`.

The confusion matrix has been prepared in the sheet `Classification (1)`. We transport the actual values of DEFAULT from the sheet `Transformed data` to the first column. This is the actual class. We copy the values predicted by the regression equation in the second column. These are the predictive scores.

Table 8.4 Coefficients table

	Coefficients	Standard error	*t* stat	*p*-value
Intercept	0.515	0.049	10.4	0.000
FEMALE	–0.049	0.010	–4.79	0.000
AGE	–0.017	0.001	–17.5	0.000
FR	0.029	0.022	1.32	0.187
GE	0.016	0.021	0.78	0.438
GR	–0.019	0.026	–0.72	0.469
HO	–0.019	0.034	–0.55	0.579
IT	–0.009	0.025	–0.36	0.721
PO	–0.024	0.038	–0.64	0.522
SP	0.005	0.023	0.21	0.833
UK	0.020	0.022	0.93	0.353
PART	–0.014	0.015	–0.91	0.363
INCOME	0.000	0.000	0.82	0.414
BALANCE	0.000	0.000	33.1	0.000

Table 8.5 Confusion matrix (cutoff 0.5)

	Actual default	Actual non-default
Predicted default	112	4
Predicted non-default	221	2,663

In the third column we calculate the predicted class. The first entry is obtained as `IF(B2 > 0.5, 1, 0)`, and, copying down, we get a predicted class for the rest of the sample. The cross-tabulation of the actual and the predicted class produces the confusion matrix of Table 8.5.

In Table 8.5, The accuracy looks like a success,

$$\text{Accuracy} = \frac{112 + 2{,}663}{3{,}000} = 92.5\%,$$

but, how useful is this model for Alexia? It performs extremely well for discarding potential non-defaulters, but not for detecting potential defaulters (112 out of 333, about one third). So, it would not be effective for default management.

So, rather than the accuracy, the relevant statistics here are the true positive and false positives rates:

$$\text{TP rate} = \frac{112}{112 + 221} = 33.6\%,$$

$$\text{FP rate} = \frac{4}{4 + 2{,}663} = 0.1\%.$$

Table 8.6 Confusion matrix (cutoff 0.4)

	Actual default	Actual non-default
Predicted default	233	44
Predicted non-default	100	2,623

The problem with this model is easy to guess: the cutoff is too high to catch the majority of the respondents. So, let us try a lower cutoff, 0.4. The new confusion matrix, calculated in the sheet `Classification (2)` of the Excel file, is Table 8.6.

The accuracy is now 95.2%. Also, we have improved the detection of the defaulters, at the price of classifying as defaulters 40 extra non-defaulters. The new statistics are:

$$\text{Accuracy} = \frac{233 + 2{,}623}{3{,}000} = 95.2\%,$$

$$\text{TP rate} = \frac{233}{233 + 100} = 70.0\%,$$

$$\text{FP rate} = \frac{44}{44 + 2{,}623} = 1.6\%.$$

Finally, we try the cutoff 0.3, which leads to the confusion matrix of Table 8.7 (sheet `Classification (3)` of the Excel file). The accuracy falls down again to 92.4%, but you may think that this model is better because of its ability to capture a higher proportion of defaulters. To be objective, we should set a **cost/benefit model** based on the estimated cost of missclassification.

$$\text{Accuracy} = \frac{293 + 2{,}480}{3{,}000} = 92.4\%.$$

$$\text{TP rate} = \frac{293}{293 + 40} = 88.0\%,$$

$$\text{FP rate} = \frac{187}{187 + 2{,}480} = 7.0\%.$$

Table 8.7 Confusion matrix (cutoff 0.3)

	Actual default	Actual non-default
Predicted default	293	187
Predicted non-default	40	2,480

8.5 Useful Tips

- Perfect prediction does not exist. Indeed, there is always a deviation between our predictions and the actual results. *You will always need to check the confusion matrix*, which contains the number of true positives (TP), false positives (FP), false negatives (FN) and true negatives (TN).
- Any model is a simplification of reality. In the example of this chapter, the classification is performed using only some basic characteristics of the client. Having a more accurate model is possible if more information is gathered, but collecting more information has a cost. *Feel comfortable with simple models and examine whether the benefit of improving the model's accuracy outweighs the costs of obtaining more information.*
- By modifying the cutoff, the number of TP's, FP's, FN's and TN's of the confusion matrix will vary, some will increase while others will decrease, always maintaining constant the column total. Whether it is better to have a low number of FN at the expense of an increase in the number of FP's, or the other way round can only be answered with a cost-benefit analysis. *Think carefully about the different cost implications.*
- What is worse, to provide a loan to someone who defaulted (FN), or not giving a loan to someone who would have paid the loan back (FP)? *Always consider the impact of FN's and FP's in your decision making.*
- *Use your experience and industry knowledge to define a good cutoff* for the classification. However, given a cost-benefit model, there are good empirical methods that can provide an optimal choice.

9 Out-of-Sample Validation

This chapter deals with the validation of classification models. The role of validation is to dismiss the concerns about overfitting, which happens when we develop a complex model in order to fit the current data but that model fails later to fit new data. The example deals with churn modeling, already mentioned in Chap. 8.

9.1 Overfitting

Overfitting is a typical problem of predictive models. Roughly speaking, overfitting occurs when a predictive model fits satisfactorily the current data but its performance is significantly worse with data which have not been used to obtain it. This typically happens with very complex models if the data set is not big enough.

The **validation** of a model is intended to convince both the developer and the user of a predictive model that there is no problem of overfitting. But it also provides a touch of professionalism which may help to build trust between the analyst and a non-technical audience.

9.2 Out-of-Sample Validation

The evaluation of a classification model is based on the confusion matrix. It may be that the statistics derived from the confusion matrix do not look so good when the model is applied to new, fresh data. In practice, the validation of a model is based on testing how the model works on data which have not been used to obtain the model. The data to which the model equation is fitted (e.g. through regression analysis) are called the **training data**, and those on which the models are tested, the **test data**.

M. Á. Canela et al., *Quantitative Methods for Management*,
https://doi.org/10.1007/978-3-030-17554-2_9

In general, users are satisfied if they can check that the performance of the model on the test set is comparable to the performance on the training set. This is called, generically, **out-of-sample validation**. The test data set is obtained in different ways, depending on the context. A simple approach is to **split** the sample in two subsamples at random, taking one subsample for training and the other subsample for testing. In the example of this chapter, we use a 50–50 split.

More sophisticated methods for validation are usually available in specialized software. For instance, in k**-fold cross-validation**, a data set is partitioned into k subsets and each of the subsets is used as a test set for a model developed on a training set resulting from merging the other $k-1$ subsets. $k = 10$ is a typical choice. This book does not cover cross-validation.

9.3 Example: The Churn Model

9.3.1 Presentation

The term churn is used in marketing to refer to a customer leaving the company. The objective of **churn modeling**, briefly discussed in Chap. 8, is to support a retention campaign by spotting potential churners.

Omicron is a young pan-European company, providing mobile phone services in the five Nordic countries. An aggressive pricing strategy has allowed Omicron to expand fast, but not all the customers are happy with the service provided by the company. The CRM director, Margrethe Thorup, considers that the **churning rate** is getting too hot. She plans to develop a model for churn prediction, which, hopefully, may allow the company to focus on some of the potential churners.

9.3.2 The Data Set

On October 7, Margrethe meets her team to discuss the project. They plan to work on data available from the Omicron's customer database. For a start, they extract a random sample from the customers database of the preceding quarter, whose accounts are still alive by September 30, monitoring them during the actual quarter. They set the sample size to 5,000.

The data set used in this first analysis (see the sheet `Data` of the Excel file `churn.xls`), which is ready at the end of the year, shows a churning rate of 19.4%. The attributes included in the data set are:

- ID, a customer ID (the phone number).
- ACLENGTH, the number of days the account had been active at the beginning of the period monitored (September 30th).
- INTPLAN, a dummy for having an international plan during at least one month of the third quarter.

- DATAPLAN, the same for a data plan.
- DATAGB, the data allowance of the data plan (either 0, 100M, 250M, 500M, 1G, 1.5G or 2G).
- OMMIN, the total minutes in calls to Omicron mobile phone numbers during the third quarter.
- OMCALL, the total number of calls to Omicron mobile phone numbers during the third quarter.
- OTMIN, the total minutes in calls to other mobile networks during the third quarter.
- OTCALL, the total number of calls to other networks during the third quarter.
- NGMIN, the total minutes in calls to nongeographic numbers, typically used by helplines and call centers, during the third quarter.
- NGCALL, the total number of calls to non-geographic numbers during the third quarter.
- IMIN, the total minutes in international calls during the third quarter.
- ICALL, the total number of international calls during the third quarter.
- CUSCALL, the total number of calls to the customer service, up to September 30th.
- CHURN, a dummy for churning during the period monitored.

A brief exploration of the data set can give us some insights. For example, the percentage of churners increases with the number of calls to customer service. This makes sense, since unsatisfied customers may be more likely to call customer service and have a higher propensity to churn.

By means of cross-tabulation or with the aid of some simple graphs, we can also detect a much higher churning rate (63%) in the subsample of customers with an international plan. As the objective of this chapter is to focus on out-of-sample validation, we will not devote more effort on data exploration. In addition, there are some variables which are redundant. We take care of them in the following subsection.

9.3.3 Dropping Redundant Information

Common sense tells us that the DATAGB should be zero when there is no data plan. This is confirmed in Table 9.1. So, there are two options to manage the data plan: (a) using the dummy DATAPLAN, or (b) using DATAGB, which means that we should include six dummies in the equation.

Table 9.1 Crosstabulation of DATAPLAN and DATAGB

	0	100M	250M	500M	1G	1.5G	2G
0	3449	0	0	0	0	0	0
1	0	74	168	291	410	522	86

The analysis would show that the choice is irrelevant, since the data plan does not contribute much to the predictive power. In our presentation, we use the dummy DATAPLAN, discarding DATAGB, so the number of terms in our equation will not exceed Excel capabilities (16 independent variables).

9.3.4 Splitting the Data Set

Our split of the data set is based on **random numbers**, that is, numbers within the 0–1 range chosen in way that the probability that the choice falls between two limits is equal to the distance between those limits. In general terms, the process goes as follows. We add a column with random numbers to the data set and sort the data based on this column. Then, we separate the fist and the second halves.

In Excel, random numbers can be obtained with the function `RAND()`. Note that, if we enter in one cell = `RAND()`, every time that we perform an action in the worksheet, the function is executed, producing a new random number. To avoid confusion, it is better to fix this, by copying the random numbers and pasting them as values on top of themselves using Excel's `Paste Special`.

More specifically, the steps are:

- Create a copy of the sheet `Data`, dropping the DATAGB column. This will be the sheet `Random ordering`.
- Insert a column at the left of the ID column. We enter = `RAND()` in all the cells of the range `A2:A5001`. This creates a column of random numbers. In `A1`, we have written the name of this column (`RANDOM`).
- With `Home` » `Copy`, copy the range `A2:A5001`.
- With `Home` » `Paste` » `Paste Values`, paste the random numbers in the same range `A2:A5001`. Now all the cells in this range have a fixed numeric value.
- Sort the range `A2:O5001` based on the column `RANDOM`.
- Split the data in two halves. In one sheet (`Training data`), put the first 2,500 customers (rows from 2 to 2,501), and in another sheet `Test data`), put the other 2,500 customers (rows from 2,502 to 5,001).

9.3.5 Regression Equation

We perform a regression analysis on the training data. The results, which can be found in the sheet `Regression`, are reported in Table 9.2. The correlation is $R = 0.465$. In spite of the sample size, some terms of the equation are not significant. Although the purpose of this example is not to test the effect of these explanatory variables, we guess that we can drop some of these terms, getting a simpler equation with similar performance. Nevertheless, we leave here the equation as it is, proceeding to perform the classification job.

Table 9.2 Regression analysis (training data)

	Coefficients	Standard Error	*t* stat	*p*-value
Intercept	–0.289	0.049	–5.90	0.000
ACLENGTH	0.000	0.000	0.10	0.923
INTPLAN	0.460	0.028	16.54	0.000
DATAPLAN	–0.015	0.015	–0.96	0.338
OMMIN	0.001	0.000	5.36	0.000
OMCALL	–0.000	0.000	–0.46	0.647
OTMIN	0.000	0.000	2.82	0.005
OTCALL	0.001	0.000	1.59	0.112
NGMIN	0.000	0.001	0.57	0.571
NGCALL	–0.000	0.002	–0.26	0.792
IMIN	0.004	0.002	2.12	0.034
ICALL	0.007	0.006	1.17	0.241
CUSCALL	0.054	0.006	9.64	0.000

9.3.6 Evaluation in the Training Set

We evaluate first our churn model in the training set as we did in Chap. 8. Since the focus of this chapter is on the validation, we skip a discussion on the choice of the cutoff. Rounding the churn rate, we get 0.19, which looks like reasonable choice (the reader can try other choices and compare). The confusion matrix is Table 9.3. In the Excel file, the results can be found in the sheet `Evaluation (training)`.

The accuracy is

$$\text{Accuracy} = \frac{331 + 1{,}448}{2{,}500} = 71.2\%.$$

The true positive rate is

$$\text{TP rate} = \frac{331}{331 + 146} = 69.4\%,$$

and the false positive rate,

$$\text{FP rate} = \frac{575}{575 + 1{,}448} = 28.4\%.$$

Table 9.3 Confusion matrix (training set)

	Actual churner	Actual non-churner
Predicted churner	331	575
Predicted non-churner	146	1,448

9.3.7 Evaluation in the Test Set

We evaluate now our churn model in the test set. Since the scores for the customers in this data set are not given by the `Analysis ToolPak`, we have to calculate them manually, writing the regression equation and applying it to the rows of the test set. In the Excel file, the coefficients of the equation are in the range `B8:B20` of the sheet `Regression` and the X values for the first customer are in the range `C2:N2` of the sheet `Test data`. So the churn score for that customer can be calculated as

```
=Regression!B$8 + Regression!B$9*'Test data'!C2 +
 Regression!B$10*'Test data'!D2 +
 Regression!B$11*'Test data'!E2 +
 Regression!B$12*'Test data'!F2 +
 Regression!B$13*'Test data'!G2 +
 Regression!B$14*'Test data'!H2 +
 Regression!B$15*'Test data'!I2 +
 Regression!B$16*'Test data'!J2 +
 Regression!B$17*'Test data'!K2 +
 Regression!B$18*'Test data'!L2 +
 Regression!B$19*'Test data'!M2 +
 Regression!B$20*'Test data'!N2
```

This and the rest of the scores can be found in column B of the sheet `Evaluation (test)`. The confusion matrix, based on the same cutoff as in Table 9.3, is Table 9.4. Now, the accuracy is 72.2%, the true positive rate is 69.7% and the false negative rate is 27.1%. In the Excel file, the results can be found in the sheet `Evaluation (test)`.

The results in the test set are quite close to those obtained for the training data. The statistics are, now:

$$\text{Accuracy} = \frac{342 + 1{,}464}{2{,}500} = 72.2\%,$$

$$\text{TP rate} = \frac{342}{342 + 149} = 69.7\%,$$

$$\text{FP rate} = \frac{545}{545 + 1{,}464} = 27.1\%.$$

Table 9.4 Confusion matrix (test set)

	Actual churner	Actual non-churner
Predicted churner	342	545
Predicted non-churner	149	1,464

So far, we have not found evidence of overfitting. In the fourth part of this book we will see other examples of out-of-sample validation, but with time series data.

9.4 Useful Tips

- Validation is not only useful in classification problems, but in all type of predictions. If the sample size allows it, *it is a good idea to divide a sample in two sets, one for training a model and the other for testing it.*
- Sometimes less is more. *Consider the use of simpler models that facilitate a wider application rather than complex models that may not work equally well in new data sets.*

Part IV
Time Series Data

Trend and Seasonality

10

In the last part of this book, we present some elementary methods to deal with time series data. This type of data has already appeared in Chaps. 1 and 2.

The example uses monthly sales data of cold weather gear. The analysis is mostly graphical, based on a linear trend with multiplicative seasonals. This provides a simple model for describing the past behavior of the data and for forecasting the values for the next year.

10.1 Time Series Data

Special methods are needed for analyzing **time series data**, as discussed in Chap. 2. Moreover, most of the methods for the analysis of time series data are specific for the **time period**. In the example of Chap. 1, the time period was the day. In Chap. 2, the minute and the hour. In this last part of the book, we deal with monthly and quarterly data. Some examples of that type of data can be:

- Monthly data on sales of a beer brand.
- Quarterly data on the gross national product (GNP) of a country.

The expression time series applies to data collected as repeated observations of the same variable at different time points. These time points are denoted by t ($t = 1, 2, 3, \ldots$). For instance, with monthly data, $t = 1$ corresponds to the first month in the data set, $t = 2$ to the second month, etc. It is easy to add a column with the t values in an Excel sheet (with the command `Fill/Series`), as we do in the example of this chapter. Figure 10.1 is a graphical representation of time series data (monthly sales) as a line plot. We put t in the horizontal axis and the data in the vertical axis.

M. Á. Canela et al., *Quantitative Methods for Management*,
https://doi.org/10.1007/978-3-030-17554-2_10

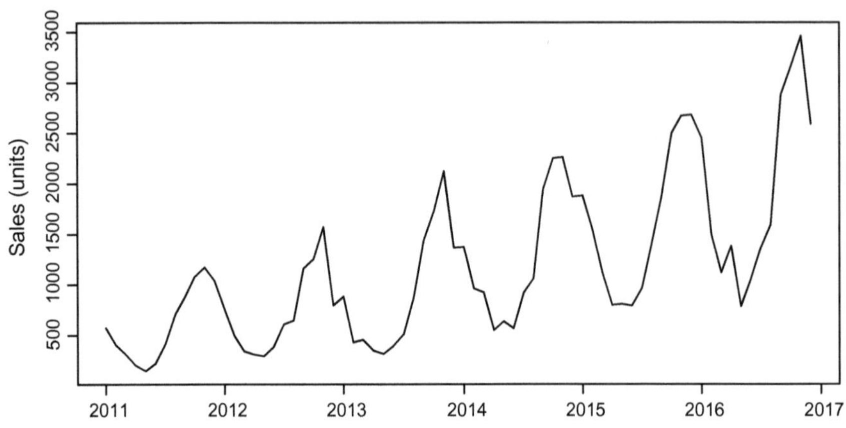

Fig. 10.1 Polar Bear monthly sales

A typical application of time series data is **forecasting** future values of the series. There are numerous methods for time series forecasting, from the simple ones presented in this book to the complex algorithms used in modern finance. The methods described here are based on modeling the data as the sum of a predictable term plus an unpredictable, random term. The predictable part is modeled in a simple way, with two components, the **trend** and the **seasonality**. Then, we use the future values of the predictable term as forecasts.

10.2 Trends

There are, essentially, two approaches for developing a trend for time series data. A **parametric trend** is given by an elementary function of t, such as $a + bt$ (linear trend), or $a + bt + ct^2$ (quadratic trend). In the example of this chapter, we use a linear trend. Having a valid parametric trend makes forecasting trivial, because the trend formula can be used for the future as well as for the past. The problem, as we discuss in many places in this part of the book, is that a trend can fit satisfactorily the current data but fail to work equally well next year.

With a **nonparametric trend**, the trend values are not calculated in terms of t by means of a fixed mathematical expression. Typically, if the actual value is available, the trend value is a linear combination of the actual and the preceding values. Then, the forecast of the future values is based on the last trend values available.

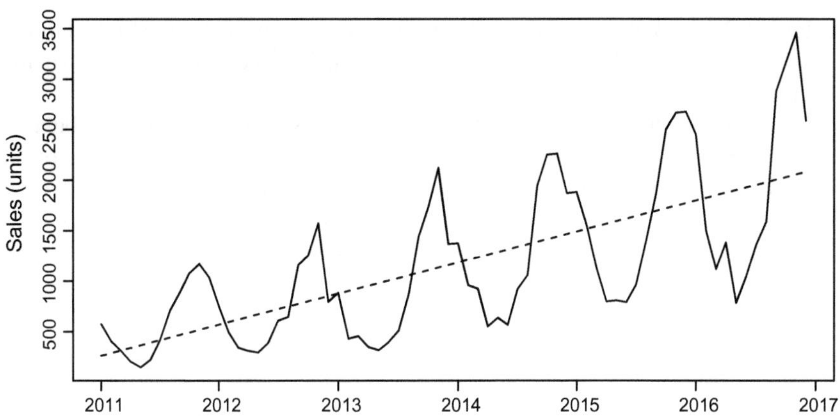

Fig. 10.2 Polar Bear data with linear trend

For a linear trend, the coefficients *a* (intercept) and *b* (slope) can be derived from a regression analysis, or obtained graphically with Excel's chart tools. In Fig. 10.2, we see again the series of Fig. 10.1, now with a linear trend superimposed.

10.3 Seasonality

The patterns that are periodically repeated in time series data are called, in a generic way, **seasonal patterns**. Seasonality is typical in monthly data (period 12 months), and also in quarterly data (period 4 months). In monthly data, seasonality is managed through 12 terms, called seasonals, which account for the "typical" evolution of the series within the year. The seasonals can be additive or multiplicative.

The **additive seasonals**, which come in the same units as the data, are summed to the trend values, improving the approximation of the actual values. The predictive model is then based on the equation

$$\text{Predicted value} = \text{Trend value} + \text{Seasonal}.$$

Additive seasonals are positive or negative, reflecting that, in some months, the actual values are above the trend and, in other months, they are below the trend.

The **multiplicative seasonals** are numbers with no units, which operate as factors. Now, the equation takes the form

$$\text{Predicted value} = \text{Trend value} \times \text{Seasonal}.$$

In the same way in which additive seasonals can be positive or negative, multiplicative seasonals can be higher or lower than one. For instance, when the seasonal factor is 1.2 for a particular month, this means that the expected value for that month is 20% above the trend. If the seasonal factor is 0.8, the expected value is 20% below the trend.

Why multiplicative seasonals? Of course, additive seasonals are simpler, but they cannot cope with a situation like that of the example, where the amplitude of the fluctuations above and below the trend, due to the seasonality, increases or decreases with the trend values. With monthly sales data, this is the rule more than the exception. The multiplicative seasonals provide a model of these fluctuations in percentage terms.

10.4 Forecasting

Forecasting (a few) future values of a series is, probably, the main application of time series analysis. In industry, for instance, sales forecasting is an essential step in production planning and inventory management. All the methods of forecasting are based on the assumption that the series is going to behave in the immediate future as it has in the recent past. Of course, this is not always true, because unexpected events can impact the mechanism that generates the series. The argument that supports forecasting methods is that these changes do not happen frequently, so the assumption is right most of the time.

These unexpected events have, in general, more impact on the trend than on the seasonals. As we will see later in this book, the nonparametric methods give more weight to recent data when developing a trend, so they are able to update the trend when the data start showing a departure from the past behavior. This makes them preferable in real industry applications such as monthly sales forecasting.

For the time interval covered by the data, the predictions can be compared to the actual data. This allows us to assess the model and, eventually, to choose between alternative models. To forecast the future values of the series, trend and seasonals have to be continued. In parametric models, it is assumed that the trend formula and the seasonals will remain valid during the time interval covered by the forecast. In nonparametric models, the forecast is based on the last available values of the trend and the seasonals.

10.5 Prediction Error

The difference of the actual value minus the predicted value is the **prediction error**,

$$\text{Prediction error} = \text{Actual value} - \text{Predicted value}.$$

When the predicted values are calculated by means of a linear regression equation, the prediction errors are called residuals, and, as we have seen in Chaps. 3 and 4, their sum is equal to zero. This is not always the case in time series analysis, in particular in the example of this chapter.

The prediction error can also be expressed in relative terms, as the percentage of the actual (or the trend) value. The analysis of the prediction error facilitates the evaluation of a forecasting model.

10.6 Example: Polar Bear Sales

10.6.1 Presentation

The Artic Experience is a company selling luxury cold weather gear, whose star product is the Polar Bear Jacket. This example (Excel file `polar.xls`) uses data on the number of units sold per month, for the period 2011–2016. The data can be found in the sheet `Data (1)`.

The line plot of Fig. 10.1, which, in the Excel file, can be found in the sheet `Line plot`, suggests the presence of an upward linear trend and a strong seasonal pattern. We explore first the trend.

10.6.2 Estimating the Trend

In Excel, trends can be obtained graphically, or extracted from a regression analysis. To add a linear trend to the line plot, as we saw in Chap. 3, we click `Add Chart Element` » `Trendline` » `Linear`. The equation can be added to the plot using `More Trendline Options`. In Fig. 10.2, we find, superimposed to the sales series, a linear trend (dashed line). In the Excel file, this plot is found in the sheet `Linear trend`.

The trend equation ($R = 0.683$) is

$$\text{SALES} = 234.5 + 25.6\,t.$$

Here, t is a time index: $t = 1$ in January 2011, $t = 2$ in February 2011, $t = 3$ in March 2011, and so on, until $t = 72$ in December 2016.

This equation has been obtained by regressing SALES on t. To do this, we put t in the first column, as in sheet `Data (2)`. The trend values can be calculated as the predicted values of the regression with the `Analysis ToolPak`. The regression report is found in the sheet `Regression analysis`.

Note. The trend equation can also be obtained graphically. But, if you use that approach, be careful, because the scale of the horizontal axis is based on dates, as in Fig. 10.2, the slope is not the same as the one reported above, for which the scale of the horizontal axis is based on $t = 1, 2, \ldots$.

10.6.3 Seasonality

Figure 10.2 shows clearly that additive seasonality would not be adequate here, since the size of the fluctuations of the actual sales around the trend increases with the trend value. So, we use multiplicative seasonals.

A simple approach (not the only one) to the estimation of the **seasonal factors** can be as follows. In the sheet `Prediction`, we have pasted the trend values besides the actual sales in column `C`, under the header `TREND`. By dividing the actual sales by the trend values, we calculate a series of multiplicative deviations in the column `D`, under the header `DEVIATION`. Then, for January, we take the mean of all the available January deviations as the seasonal factor. The same for February, March, etc.

We obtain thus 12 seasonal values: 1.386, 0.894, 0.688, 0.521, 0.429, 0.497, 0.719, 0.952, 1.443, 1.655, 1.805 and 1.343. These factors are (repeated over the years) in the column `E`, under the header `SEASONAL`. These are the factors by which we multiply the trend to obtain a prediction for the corresponding month.

Additive seasonals can be estimated in a similar way, but using additive deviations from the trend. We will see this in Chap. 11. The seasonal factors are plotted in Fig. 10.3 (sheet `Seasonals`). This graph can be regarded as the profile of a typical year.

The predicted sales are calculated in column `F` as

$$\text{PRED SALES} = \text{TREND} \times \text{SEASONAL}.$$

Figure 10.4 (see the sheet `Predicted vs. actual sales`) compares the actual sales with those predicted (dashed line).

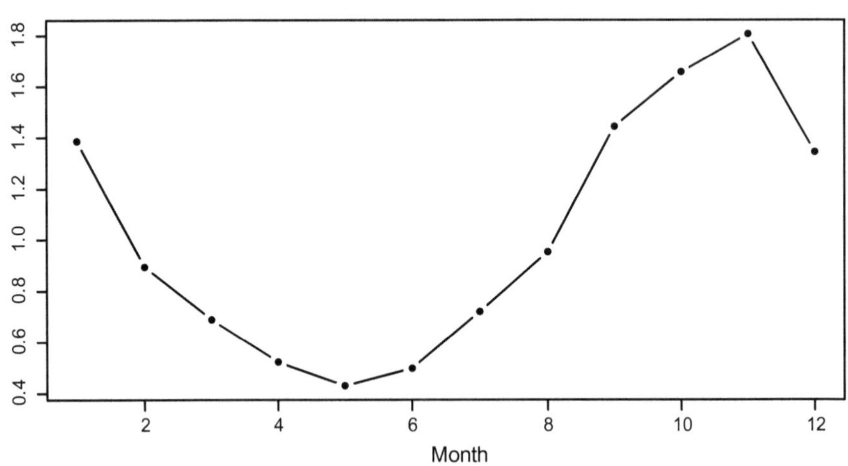

Fig. 10.3 Seasonal factors

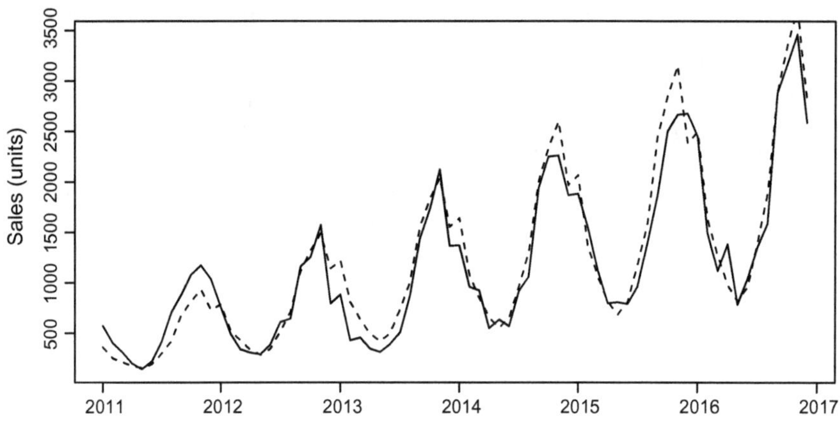

Fig. 10.4 Predicted versus actual sales

Serious textbooks prefer to use regression analysis to estimate the seasonals, but this would not change much the results, and would involve the use of logarithms. Also, some authors advise to adjust the seasonals so that the average is one.

10.6.4 Prediction Error

We calculate in column `G` of the sheet `Predictions` the prediction error as the difference of the actual sales minus the predicted sales,

$$\text{PRED ERROR} = \text{SALES} - \text{PRED SALES}.$$

The prediction errors are plotted in Fig. 10.5 (see the sheet `Prediction error`). Note that these errors are not the residuals of any regression equation and, therefore, their sum is different from zero (but not far from that). A rough estimate of the uncertainty of the forecasts can be derived from a visual inspection. A more rigorous method is based on the standard deviation of the prediction errors, which in this case is 190.3, that is, a 24% of the standard deviation of the raw sales data, which is 785.2.

10.6.5 Forecasting the Next Year Sales

Now, we can use these results to forecast the Polar Bear sales of 2017, with the formula

$$\text{FORECAST} = \text{TREND} \times \text{SEASONAL}.$$

The seasonal factors are fixed (those of Fig. 10.3), while the trend values for 2017 are calculated with the trend equation of Sect. 10.6.2. By examining the prediction error obtained for the past sales, we can guess from the data available the extent to

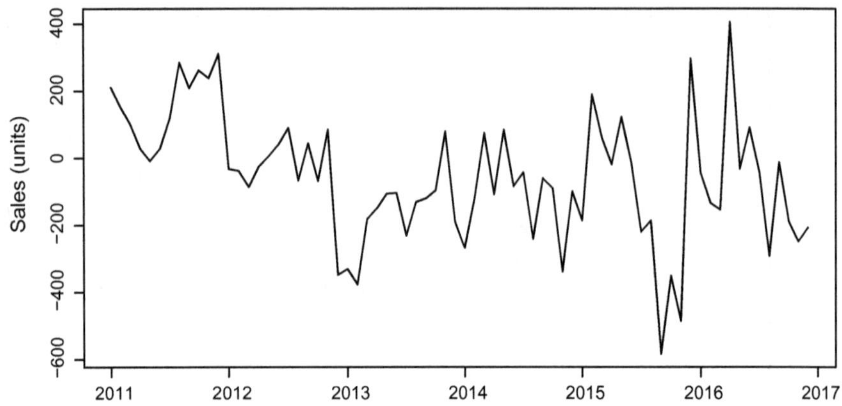

Fig. 10.5 Prediction error

Table 10.1 Forecast for 2017

t	Month	Trend	Seasonal	Forecast
73	Jan	2,106.1	1.39	2,919.7
74	Feb	2,131.7	0.89	1,904.7
75	Mar	2,157.4	0.69	1,483.3
76	Apr	2,183.0	0.52	1,138.1
77	May	2,208.6	0.43	948.4
78	Jun	2,234.3	0.50	1,109.9
79	Jul	2,259.9	0.72	1,624.5
80	Aug	2,285.6	0.95	2,176.4
81	Sep	2,311.2	1.44	3,335.1
82	Oct	2,336.8	1.66	3,867.6
83	Nov	2,362.5	1.81	4,264.7
84	Dec	2,388.1	1.34	3,207.7

which these forecasts may be wrong. These results have been calculated in the sheet `Forecast` (Table 10.1).

10.7 Useful Tips

- When analyzing time series data, *it is extremely useful to start by plotting the data.* A line plot helps to identify the trend and seasonality and to understand better the characteristics of the phenomenon under analysis.
- Expert managers tend to understand the seasonality patterns of the industry they are working at. *Seasonal patterns tend to be quite reliable over time.*

- However, *be careful with trends* since those may change abruptly, due to internal and external factors.
- In a forecast, we rely on present and past information to develop a model to predict future behavior. A forecasting model can be used to predict one period ahead, but *never use it to make long term predictions.*
- Since forecasting is difficult, *be wary of those reports which contain strong and confident statements regarding how things will be in the future.*

Nonlinear Trends

11

In many applications of time series analysis, a linear trend is not adequate for the data. In this chapter we discuss some alternatives, within the scope of Excel. In the example, we use a quadratic trend and additive seasonals for modeling the variation of monthly sales of a beer brand. We also perform an exercise of out-of-sample validation based on the last year's data.

11.1 Nonlinear Trends

A linear trend may not be adequate for describing the variation of time series data. Figure 11.1, which is a line plot of the data used in the example of this chapter, illustrates this.

Sometimes, the problem can be easily fixed by replacing the linear trend formula by another formula which provides a better description. The analyst in search of a **nonlinear trend** is typically confronted with a range of options. In Excel, for instance, to add a trend line to a chart, we can choose any of the six different trend types: Exponential, Linear, Logarithmic, Polynomial, Power and Moving Average.

The **moving average trend** is a nonparametric trend (meaning that it is not a simple function of t). We will discuss this option in Chap. 12. The other five options correspond to simple mathematical formulas. The discussion of this chapter is restricted to the **polynomial trend**.

For a polynomial trend, we have to specify the degree of the polynomial. With degree 2 (the default), we get a **quadratic trend**, whose graphical representation is a curve called **parabola**. In Fig. 11.2, we find again the data of Fig. 11.1, with both a linear trend and a quadratic trend superimposed.

The formula of the quadratic trend is $a + bt + ct^2$, where t is a time index that we already commented in Sect. 10.1. The values of the coefficients a, b and c can be obtained graphically, with the Excel's chart tools, as in the case of a linear trend

M. Á. Canela et al., *Quantitative Methods for Management*,
https://doi.org/10.1007/978-3-030-17554-2_11

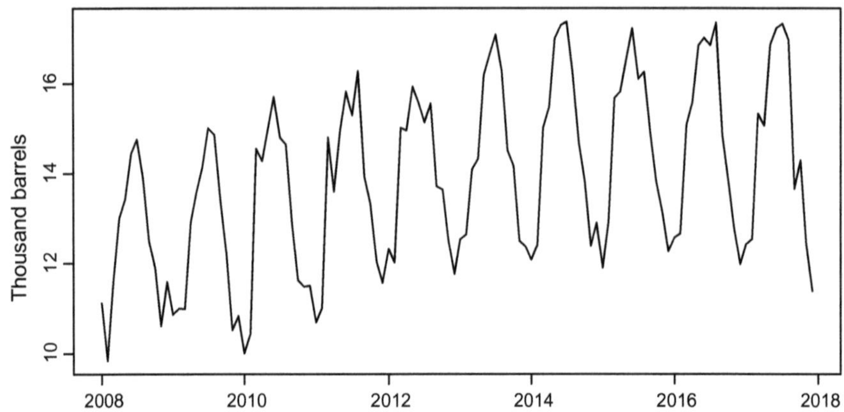

Fig. 11.1 Bayou Beer sales

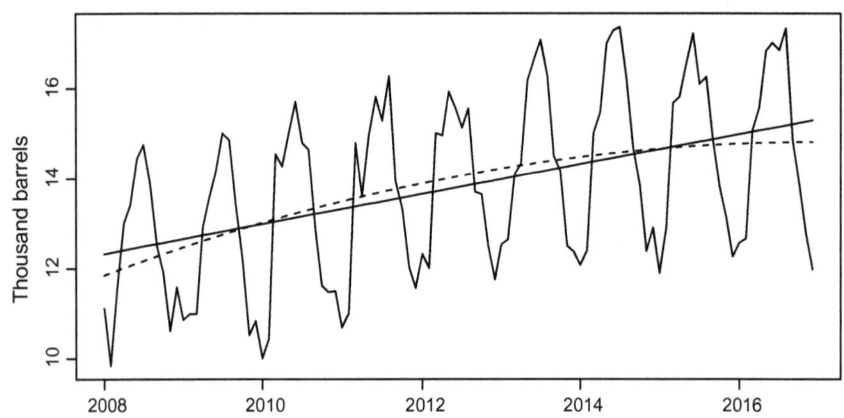

Fig. 11.2 Bayou Beer sales with linear and quadratic trends

(see Chap. 10). They can also be derived from the regression of the series data on two columns corresponding to t and t^2.

Estimating the seasonals and forecasting future values with a quadratic trend work as in the linear case. The only difference is in the way in which we obtain the trend equation. This is illustrated by the example of this chapter, in which many steps are the same as in the example of Chap. 10.

11.2 Out-of-Sample Validation

Out-of-sample validation has already been discussed in Chap. 9, where we have cautioned our readers in applying a model on new data. There, we performed a random split of the data, estimating the model equation in one half of the data and evaluating it in the other half.

For time series data, the random split does not make sense. A split based on the dates sounds much better. So, in the example of this chapter, we calculate the trend and the seasonal terms using only all the sales data except the last year. Then, we forecast the sales for that year, comparing that forecast with the actual data.

11.3 Example: The Bayou Beer Sales

11.3.1 Presentation

The Bayou is a beer brewery located in Louisiana, which was born as a taproom in New Orleans. The business of The Bayou was expanded, based on the quality of the beer served, and Bayou Beer started to be available in many bars nearby. So far, the sales of Bayou Beer are limited to South Lousiana, but their growth has been sustained during many years.

The managers at The Bayou feel that this growth is no longer going to last, and the sales trend is getting flat. They wonder about what kind of growth could be expected for the near future if they continue doing business as usual. To get inspiration, they develop a model to forecast monthly sales, one year ahead. The sales data (thousand barrels) used for the analysis cover from 2008 to 2017. In the Excel file `bayou.xls`, these data are found in the sheet `Data (1)`.

The line plot of Fig. 11.1 seems to confirm that the trend slope is decreasing. But the conclusion is still unclear, since the seasonal pattern dominates the picture.

11.3.2 Estimating the Trend

In this example, instead of using all the data available to obtain the trend and the seasonal terms, we develop a model based on the data from 2008 to 2016. Then, we use the model to forecast the sales of 2017. So, we have nine years of data for **training** the model and one year of data for **testing** it.

The training data are in the sheet `Data (2)`. Regressing the sales on the time index t, we get the trend equation

$$\text{SALES} = 12.299 + 0.0277\,t.$$

The correlation is $R = 0.443$. Here, t goes from $t = 1$ in January 2008 to $t = 108$ in December 2016. To estimate the quadratic trend equation, we add a column with the t^2 values, from $t^2 = 1$ to $t^2 = 108^2 = 11,664$, regressing SALES on t and t^2. In the Excel file, the sheet `Data (3)` contains the two columns needed for getting these results in the `Analysis ToolPak`.

The quadratic equation is

$$\text{SALES} = 11.797 + 0.055\,t - 0.0003\,t^2.$$

The correlation increases to $R = 0.457$. In Fig. 11.2 we see, superimposed to the sales series, a linear trend (solid line) and quadratic trend (dashed line). These trends can be produced with the Excel's chart tools.

Of course, the quadratic equation fits better the data than the linear equation, but is the improvement relevant? We could be tempted to answer this question based on the nonsignificance of the quadratic term in the enlarged equation (see the regression report in the sheet `Quadratic regression` for that), but we prefer to look at Fig. 11.2, which suggests that the quadratic trend can be better for forecasting purposes. Note that, with the continuation of the linear trend to 2017, the predictions could be too optimistic, if the trend were getting flat. On the other way, the quadratic trend goes downwards, so it could lead to a pessimistic forecast.

11.3.3 Seasonality

Figure 11.2 shows that additive seasonality can be adequate here, since the size of the fluctuations around the trend is more or less constant across years. So, for a change, we use **additive seasonals** in this example, not meaning that multiplicative seasonals would not be adequate. All the calculations are in the sheet `Prediction`.

As we did for the multiplicative seasonals in the preceding chapter, we follow a simple approach to the estimation of the seasonals here. By subtracting from the actual sales the trend values, we get a series of additive deviations (the column `DEVIATION` in the sheet `Prediction`). Then, for January, we take the mean of all the available January deviations as the seasonal term. The same for February, March, etc. We obtain thus the 12 seasonal terms: –2.087, –2.024, 0.373, 0.696, 1.727, 2.189, 1.993, 1.852, 0.054, –0.751, –1.933 and –2.091 (in the column `SEASONAL`). These are the terms that we sum to the trend values to obtain the predicted sales for the corresponding month.

The seasonals are plotted in Fig. 11.3 (in the Excel file, in the sheet `Seasonals`) which can be regarded as the profile of a typical year. Figure 11.4 compares the actual sales (solid line) with the predicted sales (dashed line). The calculation of the predicted sales can be found in the column `PRED SALES` of the sheet `Prediction`, as

$$\text{PRED SALES} = \text{TREND} + \text{SEASONAL}.$$

Some textbooks prefer to use regression analysis to estimate the seasonals, adding dummies for the months (11 dummies), which does not change much the results. Also, some authors advise to adjust the seasonals so that the average is zero.

11.3.4 Prediction Error

We calculate, in the column `PRED ERROR` of the same sheet, the prediction error as the difference of the actual sales minus the predicted sales,

$$\text{PRED ERROR} = \text{SALES} - \text{PRED SALES}.$$

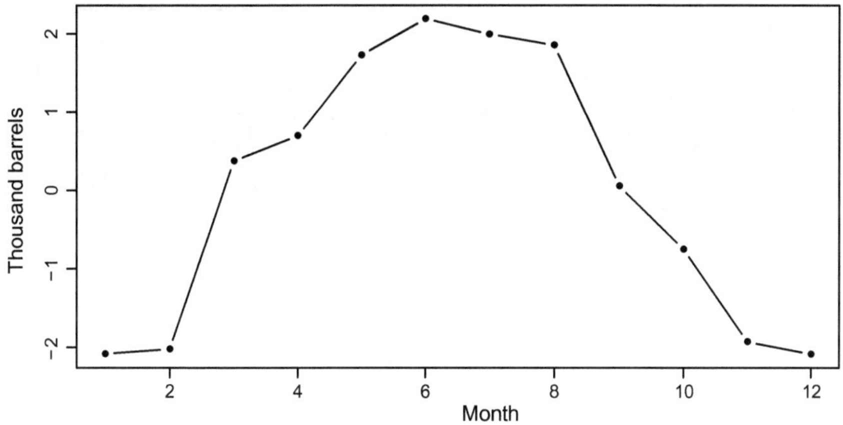

Fig. 11.3 Seasonals

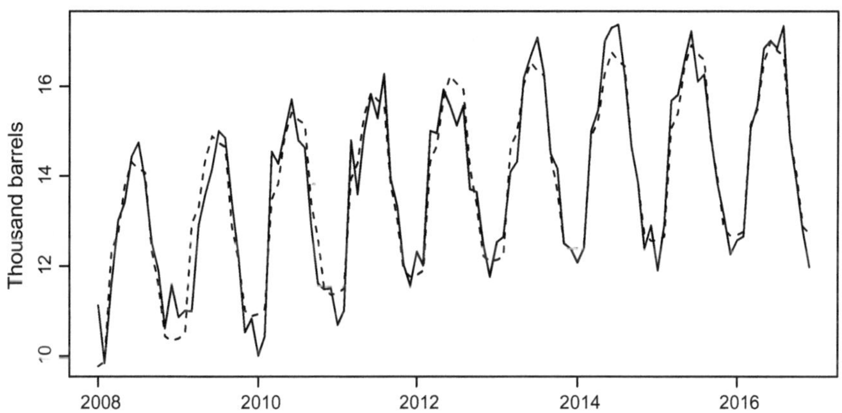

Fig. 11.4 Predicted versus actual sales

The prediction error is plotted in Fig. 11.5 (in the Excel file, in the sheet `Prediction error`).

Note that these errors are not the residuals of any regression equation (they would be if we were using dummies) and, therefore, their sum is different from zero. A rough estimate of the uncertainty of the forecasts can be derived from a visual inspection. A more objective method is based on the residual standard deviation. The standard deviation of the prediction error is 0.513, that is, a 26% of the standard deviation of the actual sales, which is 1.958.

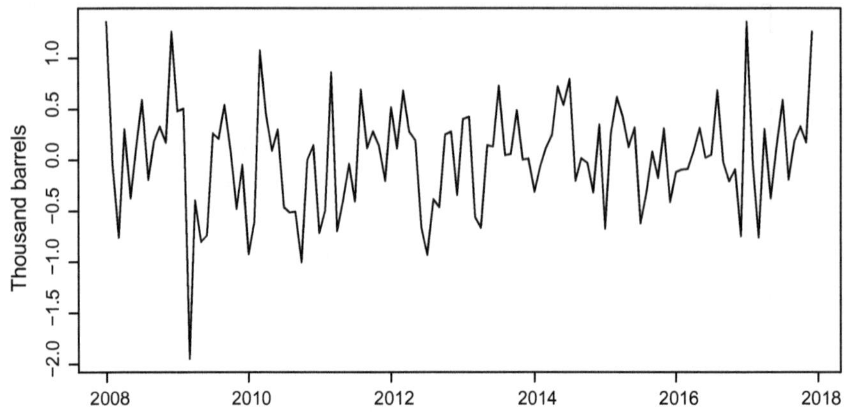

Fig. 11.5 Prediction error

11.3.5 Forecasting the Sales for the Year 2017

Now, we can use these results to forecast the Bayou Beer sales of 2017, with the formula

$$\text{FORECAST} = \text{TREND} + \text{SEASONAL}.$$

The seasonal factors are fixed (those of Fig. 11.3), and the trend values for 2017 are calculated with the quadratic trend equation given above, using $t = 109, \ldots, 120$.

The results are presented in Table 11.1. In most cases, the forecasting error is quite small. So, we can conclude that the continuation of the quadratic trend works in this

Table 11.1 Forecast for 2017

Month	Trend	Seasonal	Forecast	Actual sales	Error
Jan	14.817	–2.087	12.731	12.421	–0.310
Feb	14.817	–2.024	12.794	12.544	–0.250
Mar	14.817	0.373	15.190	15.324	0.134
Apr	14.816	0.696	15.512	15.063	–0.449
May	14.815	1.727	16.542	16.866	0.324
Jun	14.813	2.189	17.002	17.230	0.228
Jul	14.810	1.993	16.804	17.329	0.525
Aug	14.807	1.852	16.660	16.965	0.305
Sep	14.804	0.054	14.858	13.658	−1.200
Oct	14.800	–0.751	14.049	14.293	0.244
Nov	14.796	–1.933	12.863	12.404	–0.459
Dec	14.791	–2.091	12.699	11.382	–1.317

case. Note that our forecasting method works better for the first months of the year, getting worse as time runs. This typically happens when forecasting one year ahead. So, practitioners use such forecasts for planning the next year, adjusting the forecast every month.

11.4 Useful Tips

- If a series (i.e. monthly sales) follows a trend given by a quadratic formula $a + bt + ct^2$, understanding how the series changes with time is relatively easy in a chart such as Fig. 11.2. But *the interpretation of the coefficients of t and t^2 is not easy.*
- It is unrealistic to expect a trend to last long. Trends change, so *you must update your model with the new observations.*
- Parametric trends put the same weight on events that happened long ago as on recent events. Therefore, in most cases, *it is recommended to drop very old data, since they can bias the current forecast.*
- With a line plot, you can visually check whether the amplitude of the fluctuations above and below the trend is constant over time. *If it is, you can choose between additive and multiplicative seasonals. If it is not, better use multiplicative seasonals.*

Moving Average Trends 12

This chapter deals with time series trends based on moving average formulas. A special case, in which the trend values are calculated with the exponential smoothing formula, is discussed with more detail.

The example uses data on quarterly sales of brandy in Australia. The analysis presented illustrates the application of moving average methods with both raw and deseasonalized data.

12.1 Nonparametric Trends

In the last two chapters of this book, which are a bit more mathematical than the rest, we discuss some simple methods to obtain **nonparametric trends**, and the advantages and disadvantages of the nonparametric approach. While this chapter is rather general, Chap. 13 is devoted to a specific technique, the Holt-Winters forecasting method, which is the favorite among industry practitioners.

In nonparametric trends, the trend value is not obtained by means of a fixed function of t, but as the (weighted) average of values of the series which are close in time to t. Besides the parametric trends discussed in Chaps. 11 and 12, Excel provides two nonparametric trend methods, the moving average and the exponential smoothing.

In certain applications, typically in operations management, the nonparametric trend is presented as a new series, derived from the original one by means of a **smoothing method**. This smoothed series is taken as a better representation of the real phenomenon which generates the data, because the smoothing procedure (a moving average formula in most cases) is thought to remove part of the "noise". For instance, this is useful when dealing with industrial production data which is affected by measurement error, fluctuations in the quality of the raw materials, etc.

M. Á. Canela et al., *Quantitative Methods for Management*,
https://doi.org/10.1007/978-3-030-17554-2_12

12.2 Moving Average

Most methods to obtain nonparametric trends are based on a **moving average** (MA) formula. A moving average trend value at t is obtained as a (weighted) average of a collection values of the series which are picked around t. Usually, the weights are all equal.

To be more specific, let us denote by $x(t)$ the value of the series at time t. An example of a uniformly weighted MA formula would be:

$$\mathrm{ma}(t) = (1/4)\big[x(t) + x(t-1) + x(t-2) + x(t-3)\big]. \tag{12.1}$$

Sometimes the average is centered in the actual observation, as in

$$\mathrm{ma}(t) = (1/3)\big[x(t-1) + x(t) + x(t+1)\big]. \tag{12.2}$$

Both formulas can be used for a description of the past behavior of the series. Although formula 12.2 is more elegant, formula 12.1 allows us to carry the trend as far as the data.

The choice of the number of terms involved in the moving average is based on the nature of the data. With quarterly data, as in the example, it would be natural to take observations from the four quarters. To center the average in the actual observation, we can organize that as

$$\mathrm{ma}(t) = (1/8)x(t-2) + (1/4)\big[x(t-1) + x(t) + x(t+1)\big] + (1/8)x(t+2)\big]. \tag{12.3}$$

Here, to keep the symmetry, we have halved the weight at the extremes. We use formula 12.3 in the example of this chapter, in which we deal with quarterly sales data. Any of these three MA formulas produces a new series which is smoother than the original series.

In the Excel trendline options, the option `Moving Average` is based on a uniformly weighted average. The user sets the parameter `Period`, which is the number of terms averaged. With period 4, this corresponds to formula 12.1.

12.3 Exponential Smoothing

By means of the **exponential smoothing** formula, we derive from the original series $x(t)$ a new, smoother series, $\mathrm{sm}(t)$. The value of the smoothed series at time t is a weighted average of the current value of the original series $x(t)$ and the preceding value of the smoothed series $\mathrm{sm}(t-1)$,

$$\mathrm{sm}(t) = \alpha\, x(t) + (1-\alpha)\mathrm{sm}(t-1). \tag{12.4}$$

α is called the **smoothing parameter** ($0 < \alpha < 1$). The exponentially smoothed series can be used as a nonparametric trend. In that case, a low value of α is used. A typical choice is 0.2.

In a trend obtained through exponential smoothing, every time that we have a new data point, the trend is updated as the weighted average of the former trend value and the new observation. The parameter α is the weight given to the new observation. Figure 12.1 illustrates this.

The four data points in Fig. 12.1 are the first observations of the series of the example of this chapter (solid line). The dashed line corresponds to a exponential smoothing. On the left side, by using $\alpha = 0.2$, we get a smoother trend, since the trend is less sensible to the fluctuations of the current observation. On the right side, with $\alpha = 0.7$, the trend captures most of the variation of the series.

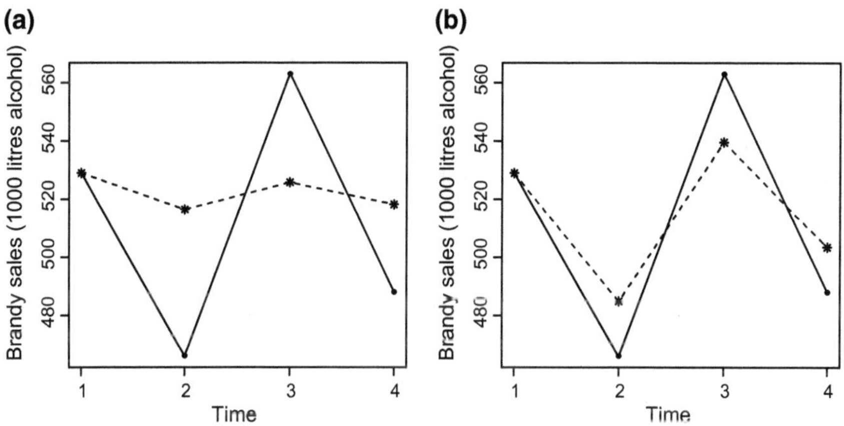

Fig. 12.1 **a** Smoothing (alpha = 0.2) **b** Smoothing (alpha = 0.7)

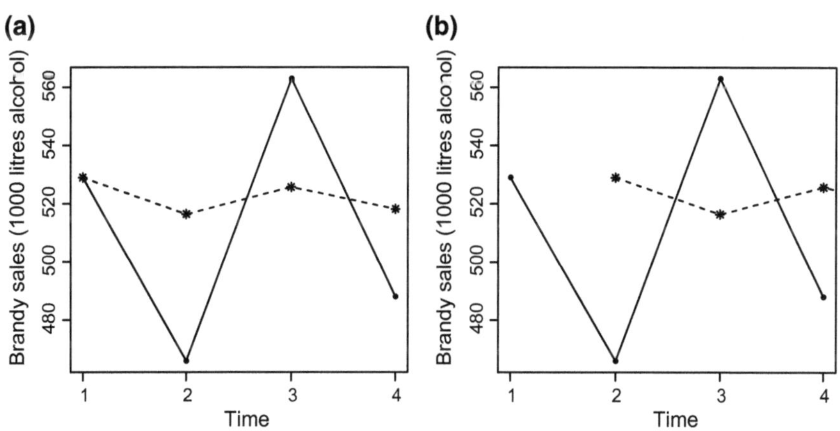

Fig. 12.2 **a** Smoothing as a trend **b** Smoothing as a forecast

Exponential smoothing can also be used in forecasting one-period ahead. We take $\mathrm{sm}(t)$ as the forecast for time $t + 1$. Graphically, this means shifting the smoothed line to the right, as shown in Fig. 12.2, which uses the same data as Fig. 12.1.

In Excel, exponential smoothing is suggested as a forecasting method within the `Analysis Toolpak`. $1 - \alpha$ is called there the **damping factor**. The default for the damping factor in Excel is 0.3 (which is equivalent to $\alpha = 0.7$).

Note. Exponential smoothing is sometimes called exponentially weighted moving average (EWMA), since its formula can be rewritten as a moving average formula in which the weights of the observations decrease exponentially as they get older.

12.4 Deseasonalizing

The data of Figs. 12.1 and 12.2, which belong to the series used in the example, are quarterly sales, showing a seasonal pattern (summer, autumn, winter, and spring). In Fig. 12.2, it is clear that the exponential smoothing formula does not anticipate the next seasonal variation.

When there are seasonal patterns, typically with quarterly and monthly data, the seasonal variation can interfere in the calculation of an exponential smoothing forecast. So, we may discount the seasonal effects before applying the exponential smoother.

A series resulting from discounting the seasonal effects from the original series is said to be **deseasonalized**. There are many methods to deseasonalize. In the example, we use a simple approach, based on multiplicative seasonals, available from a previous analysis. The deseasonalization is performed by dividing the actual observation by the seasonal.

12.5 Example: Brandy Consumption in Australia

12.5.1 Presentation

We are in mid 2014. Brandy sales in Australia have been declining for years, but there is a rumor about a recovery in late 2014 and 2015. Jackson, a wine merchant, decides to collect brandy sales data from public sources to analyse them in order to have a more informed picture about brandy sales.

Jackson's analyst takes a look at the quarterly data on sales of brandy in Australia, available online from 1985 to mid 2014, from governmental sources. Based on these data, he wants to develop a model to forecast the sales of next quarter. If the model is accurate, his plan is to update it every new quarter and so be able to better adapt Jackson's business to the market evolution.

12.5.2 Parametric Trend Approach

The data for this example, extracted from www.abs.gov.au, can be found in the first sheet (`Data (1)`) of the Excel file `brandy.xls`. They are quarterly sales of brandy (1000 l of alcohol) in Australia, covering from the first quarter of 1985 to the second quarter of 2014.

Our analysis, tales the same perspective as in Chap. 11. We leave aside a segment of the data, which is used to evaluate the forecast. Here, we place ourselves at the beginning of 2014, as if the last two observations were not available. These two observations are included in Fig. 12.3, but not if in Figs. 12.4, 12.5, 12.6 and 12.7.

The line plot of Fig. 12.3 shows the decrease in brandy sales. But the conclusion about the potential recovery is unclear from this plot, given the strong seasonal pattern shown by the data.

The linear trend equation ($R = 0.905$) is

$$\text{SALES} = 456.1 - 3.62\,t.$$

Here, t is a time index, going from $t = 1$ in the first quarter of 1985 to $t = 116$ in the last quarter of 2013. Figure 12.4 shows, superimposed to the sales series, the linear and the quadratic trends, which can be easily produced with the Excel's chart tools.

We see clearly in Fig. 12.4 that the linear trend does not account for the behavior of the sales during the last ten years. In spite of the strong correlation, the linear trend would not be useful for forecasting purposes. The fit would improve leaving out the first fifteen or twenty years, but the length of the series would be dramatically reduced.

The quadratic trend ($R = 0.926$) looks more appropriate in the figure. The equation is

$$\text{SALES} = 516.6 - 6.70\,t + 0.026\,t^2.$$

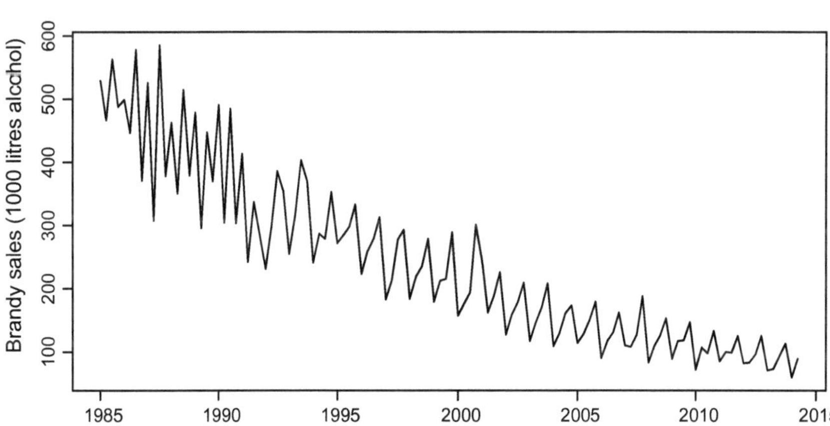

Fig. 12.3 Brandy quarterly sales

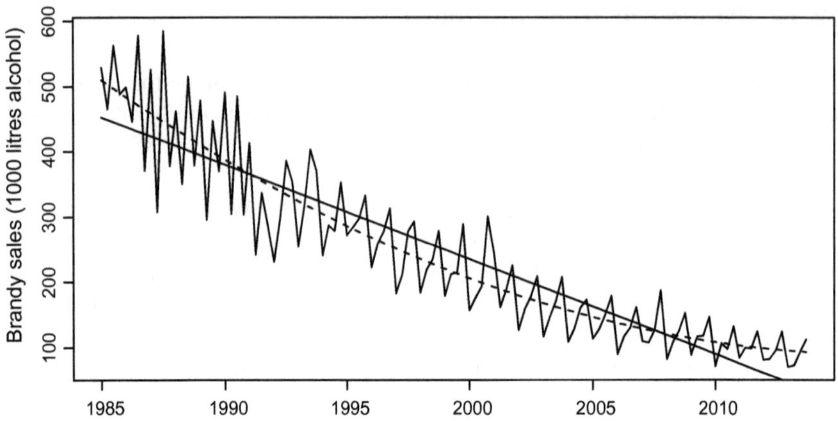

Fig. 12.4 Brandy sales with linear and quadratic trends

12.5.3 Moving Average Trend

Figure 12.5 shows again the data, now with a superimposed MA trend. This trend has been obtained by applying directly the formula 12.3 of Sect. 12.2. We have not used the `Analysis ToolPak`, since the `Moving Average` option, available there, is restricted to uniformly weighted moving averages. The calculations can be found in the sheet `MA calculation`. Note that the MA trend has four data points less than the original series. Indeed, the MA values for the first quarters of 1985 and the last two quarters of 2013 cannot be calculated.

Looking at the trend of Fig. 12.5, it seems that, except for the two humps in 1992 and 2001, a description in terms of linear trends is acceptable, if we just allow the

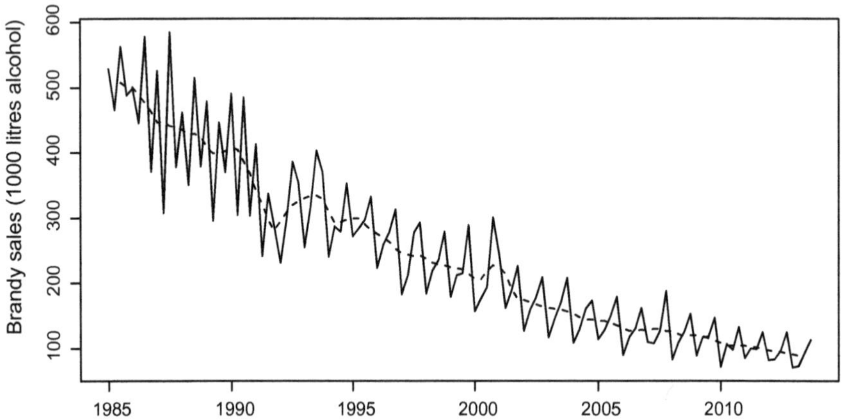

Fig. 12.5 Brandy sales with MA trend

slope to change in 2000s. Also, it does not seem that the trend is yet flat. The quadratic trend seems appropriate for a description of the sales in the years 1985–2013, but not for forecasting 2014, since it may be too optimistic. Of course, the trend of Fig. 12.5 is only good for understanding the historical data, not for forecasting. But it helps to guess what may go wrong in predicting future sales.

12.5.4 Seasonality

As we have said before in this chapter, deseasonalizing before applying the exponential smoother is recommended when the seasonal variation is strong, as in this example. We use multiplicative seasonals here.

We have seen in Chap. 10 how to calculate seasonal factors by averaging the deviations of the actual sales from the trend values. We use here the quadratic trend, to have 29 complete years, but the results obtained with the MA trend would not be much different.

By dividing the actual sales by the trend we get a multiplicative deviation for every data point. In the Excel file, the calculations can be found in the sheet `Seasonality`. The deviations for the four quarters of 1985 are 1.037, 0.926, 1.133 and 0.995, respectively. The interpretation is as in Chap. 10: in the first quarter, sales were 3.7% above trend, in the second quarter, 7.4% below trend, etc. For 2013, the four deviations are 0.741, 0.767, 0.995 and 1.204. Contrary to our expectations, the seasonal pattern seems to have changed since 1985, and a set of seasonals for the whole period covered by the data does not make sense.

Indeed, at the beginning of the series the pattern gives higher sales in the first and third quarter, while, at the end, we have high sales in the fourth quarter (Christmas?) and low sales in the first quarter. Since we wish to forecast the sales in 2014, we need reliable seasonal factors for the last part of the series. So, we restrict the last part of the discussion to the period 2001–2013.

12.5.5 Deseasonalized Sales

In the Excel file, we have packed in the sheet `Exponential smoothing` the rest of the calculations for this example. The data (52 rows) cover only the 13 years that were kept. The columns `TREND` and `DEVIATION` contain the same data as in the sheet `Seasonality`. The seasonal factors of the column `SEASONAL` are calculated as in Chap. 10: the factor for a specific quarter is the average of the deviations calculated in the previous column for that quarter.

The seasonals factors thus obtained are 0.781, 0.893, 1.017 and 1.279. The next column, `DESEASONALIZED`, contains the deseasonalized sales, calculated as the ratio of the sales by the seasonal factors. We see in Fig. 12.6 both the actual sales (solid line) and the deseasonalized sales (dashed line).

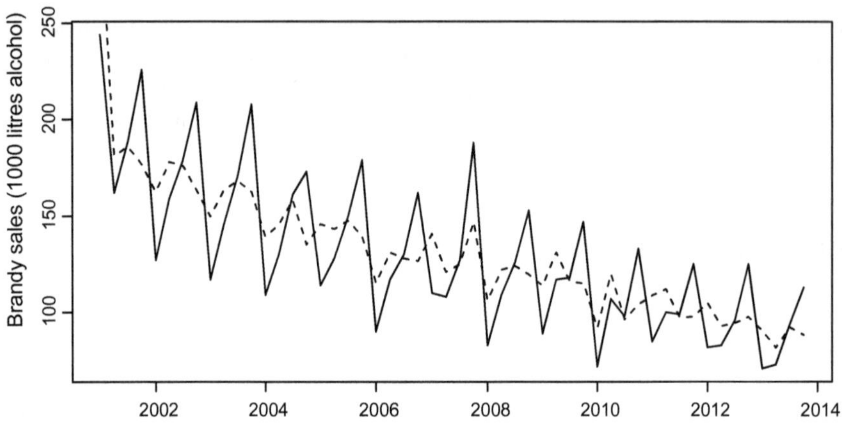

Fig. 12.6 Deseasonalized brandy sales

12.5.6 Exponential Smoothing

Finally, we apply the exponential smoother to the deseasonalized sales, using $\alpha = 0.2$. The smoothed series is in the column `EXP SMOOTHING`. Figure 12.7 shows the actual sales and the outcome of the exponential smoothing.

Now, the smoothed series is taken as the trend. To forecast the first quarter of 2014, we multiply the last trend value available (`G53`) by the seasonal factor of the first quarter (`E2`),

$$93.5 \times 0.781 = 73.0.$$

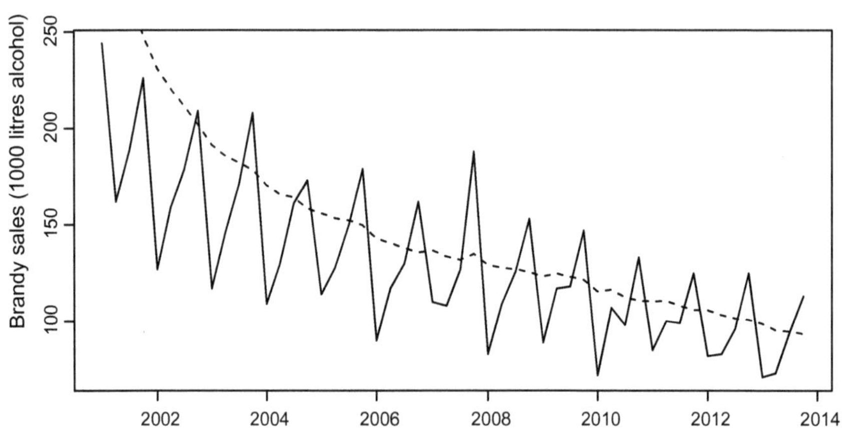

Fig. 12.7 Brandy sales with exponential smoothing trend

The same for the second quarter,

$$93.5 \times 0.893 = 83.5.$$

The actual values for these two quarters are 60 and 89. In normal practice, the exponential smoothed series would have been updated after getting the sales of the first quarter, so the forecast for the second quarter would have been a bit different.

12.6 Useful Tips

- Forecasting methods are not effective for long term forecasting (i.e. sales five years from now). *Use your forecasting model only to predict the following period and recalibrate it once you have new observations.*
- *If the slope of the trend can change unexpectedly, better use a nonparametric trend than a parametric trend.* The non-parametric trend captures faster the changes.
- Alpha (α) is a parameter that defines the weight the nonparametric trend allocates to the current observation. $\alpha = 0.2$ allocates 20% of the weight to the current observation and 80% to past observations. *If the data you are analyzing are erratic, a small alpha will help you eliminate part of the noise that may mislead your forecast.* In business, $\alpha = 0.2$ is fairly common.

13 Holt-Winters Forecasting

The last chapter of this book presents the Holt-Winters forecasting method, in two versions, with additive and with multiplicative seasonals. The example is an application of the Holt-Winters method with multiplicative seasonals to data on monthly sales of a meat product at an Ecuadorian supermarket chain.

13.1 Introduction

This chapter is more technical than the rest of the book. It describes the **Holt-Winters forecasting method** for monthly time series data, which is the favorite of practitioners. The Holt-Winters method uses a nonparametric trend, updating both the trend and the seasonals by means of a combination of three exponential smoothing formulas. The name **triple exponential smoothing** (ETS) is also used for this method.

Although the idea behind the method is easy to grasp, Holt-Winters formulas may look too involved to you. Nevertheless, with the use of analytical software, Holt-Winters forecasts are easy to obtain. Therefore, you can leave aside the formulas in a first lecture, focusing on a general understanding of the method.

Holt-Winters forecasting is available in the computer from many sources, including some add-ins for Excel, with slight differences in the formulas used. There is one version with additive seasonals and another version with multiplicative seasonals (the one used in the example of this chapter). There is also a version with a multiplicative slope, and versions for quarterly data. The Excel function `FORECAST.ETS` (not available in Excel for Macintosh) implements the formulas of Sect. 13.6.

The example of this chapter shows that, although the formulas may look a bit complex, they are simple enough to be implemented in an Excel sheet without much pain. The companion Excel file includes all the calculations for the data of the example. These sheets can be adapted to other data sets, so you can easily develop a template for your data.

M. Á. Canela et al., *Quantitative Methods for Management*,
https://doi.org/10.1007/978-3-030-17554-2_13

13.2 The Holt-Winters Approach

In the Holt-Winters method, we model the data, as usual, combining a trend with seasonal patterns. The trend can be seen as a linear trend in which the slope is updated every time that a new observation is available. The seasonals, either additive or multiplicative, are updated as well. The updates are based on exponential smoothing formulas.

The model is usually presented as based on three components, the **trend**, the **slope** and the **seasonal** components. Roughly speaking, the trend and the seasonal are used to calculate the predicted values, while the slope is used to update the trend. More specifically, the slope is interpreted as the change between consecutive trend values. The inspiration for this comes from the fact that, for a linear trend $a + b\,t$, the difference between two consecutive trend values is equal to the slope b.

For each component, there is a parameter used to weight the new observation when updating. These parameters, denoted by α, β and γ, are called the **smoothing parameters**. The choices $\alpha = \beta = \gamma = 0.2$ are typical. Nevertheless, we try a variation ($\alpha = 0.5$) in the example.

In the Holt-Winters model, the slope is not fixed, but changes across months. It may be the case that the slope is positive in one part of the series but negative in another part. When we arrive at the end of the actual data, the slope is no longer updated, so that in forecasting the future values we use the last slope available. The seasonals, either additive or multiplicative, are also continuously updated. The last values available are those used to predict future values.

Note. The Holt-Winters trend is usually called level, but we have preferred to keep the term trend here, to make it simpler. Also, "level" may be wrongly understood as "intercept", that is, as the constant term of the linear trend formula.

13.3 Multiplicative Seasonals

The multiplicative Holt-Winters model is based on a TREND × SEASONAL approach. To show the details, let us denote the length of the series by N, the series by $x(t)$, the trend by $a(t)$, the slope by $b(t)$ and the seasonal by $s(t)$. The basic formulas are:

$$a(t) = \alpha\,\frac{x(t)}{s(t-12)} + (1-\alpha)\big[a(t-1) + b(t-1)\big], \tag{13.1}$$

$$b(t) = \beta\big[a(t) - a(t-1)\big] + (1-\beta)\,b(t-1), \tag{13.2}$$

$$s(t) = \gamma\,\frac{x(t)}{a(t)} + (1-\gamma)\,s(t-12). \tag{13.3}$$

Each of these three formulas is based on a weighted average of the current and the preceding observations. For instance, let us take $\alpha = 0.2$ in formula 13.1. Then, the trend value is obtained as the weighted average of: (a) the result of updating the previous trend value by summing the slope (weight 80%), and (b) the new observation after discounting the seasonal pattern (weight 20%).

With $\beta = 0.2$, formula 13.2 updates the slope as the weighted average of the previous slope (weight 80%) and the current change in the trend (weight 20%). A similar idea is applied to the seasonals in formula 13.3.

13.4 The First Year

For the first year, since there are not enough data available, formulas 13.1–13.3 are slightly changed. First, the subscript $t - 12$ does not make sense in formulas 13.1 and 13.3. Also, the subscript $t - 1$ does not make sense in any of the formulas for the first January. There are many ways to fix this, but the method is not relevant for the forecaster, because the model adjusts itself as time runs. Here, we have chosen a simple option.

For the seasonals, the best values available are used in the first year. When there are no previous estimates, it is recommended to set the first 12 seasonals to one, as we do in the example. This has little effect on the final forecasts, as you may check by trying other options in the Excel file of the example. The Holt-Winters formulas correct the wrong initial estimates of the seasonals.

Then, the January slope is set to zero and the trend to the first observation minus the seasonal ($a(1) = x(1)/s(1), b(1) = 0$). For the months February–December, the slope is calculated with formula 13.2, and the trend with

$$a(t) = \alpha \frac{x(t)}{s(t)} + (1 - \alpha)\big[a(t - 1) + b(t - 1)\big]. \tag{13.4}$$

13.5 Forecasting One Year Ahead

We also have to adapt formulas 13.1–13.3 to forecast the values of the series one-year ahead. Since there are no available actual values $x(t)$, the weighted averages no longer make sense. So, we take $\alpha = \beta = \gamma = 0$, getting

$$a(t) = a(t - 1) + b(t - 1), \tag{13.5}$$

$$b(t) = b(t - 1), \tag{13.6}$$

$$s(t) = s(t - 12). \tag{13.7}$$

13.6 Additive Seasonals

The additive Holt-Winters model is based on a TREND + SEASONAL approach. The formulas for the trend, the slope and the seasonal components are similar to those of the multiplicative model, but replacing the two quotients by differences. We rewrite them completely below.

The basic formulas become:

$$a(t) = \alpha\big[x(t) - s(t-12)\big] + (1-\alpha)\big[a(t-1) + b(t-1)\big], \tag{13.8}$$

$$b(t) = \beta\big[a(t) - a(t-1)\big] + (1-\beta)\,b(t-1), \tag{13.9}$$

$$s(t) = \gamma\big[x(t) - a(t)\big] + (1-\gamma)\,s(t-12). \tag{13.10}$$

For the first year, in which formulas 13.8–13.10 cannot yet be applied, the solutions are similar to those recommended in the multiplicative case. If no seasonal estimates are available, the seasonals can be set to zero for the first year. We start with $a_1 = x_1 - s_1$, $b_1 = 0$. For the months February–December, the slope is calculated with formula 13.9, and the trend with

$$a(t) = \alpha\big[x(t) - s(t)\big] + (1-\alpha)\big[a(t-1) + b(t-1)\big]. \tag{13.11}$$

Finally, to forecast the series values one-year ahead, we use:

$$a(t) = a(t-1) + b(t-1), \tag{13.12}$$

$$b(t) = b(t-1), \tag{13.13}$$

$$s(t) = s(t-12). \tag{13.14}$$

13.7 Example: Supermercados Andinos

13.7.1 Presentation

Supermercados Andinos is an Ecuadorian supermarket chain, which has been selling food through Internet during the last years. This example uses data on the monthly sales of a product of broad diffusion, the Rib Eye steaks of Angus beef, which are delivered in packages of two. Unpredictable fluctuations in the sales of this product make forecasting a challenging.

The Excel file `andinos.xls` contains data on monthly unit sales, from January 2013 to June 2017 (sheet `Data`). Figure 13.1 is the corresponding line plot. It does not show any clear trend, and the potential seasonality patterns are unclear.

We use the multiplicative Holt-Winters method to develop a forecasting model, with a focus on short term forecast. To develop the model, we use the data from

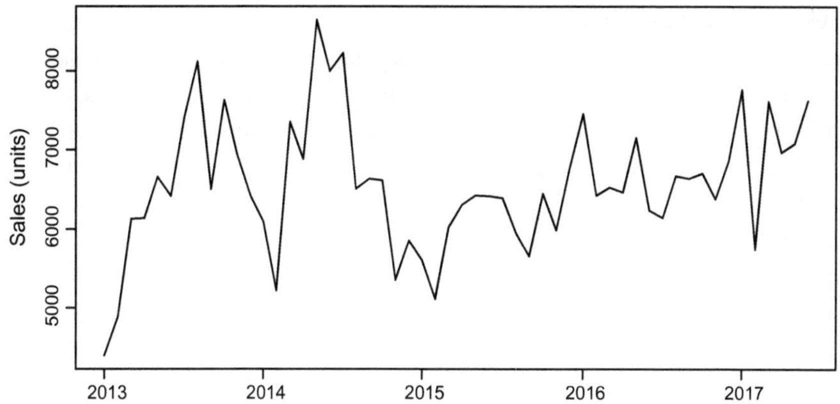

Fig. 13.1 Rib Eye monthly sales

the years 2013–2016. Then, we forecast the sales for the first six months of 2017, comparing the actual sales with our prediction. We start with $\alpha = \beta = \gamma = 0.2$, trying later an alternative value for α.

The sheet `HW (alpha = 0.2)` contains all the calculations for our first Holt-Winters model. We discuss the detail in this and the three following subsections.

13.7.2 Initializing the Model

As explained in Sect. 13.4, we have to initialize the Holt-Winters model by providing values for the seasonals of the first year and the slope of the first month. As suggested there, we set the seasonals as one ($s(1) = \cdots = s(12) = 1$), so the first trend value is equal to the sales in January 2013, ($a(1) = x(1)$), and the initial slope as zero ($b(1) = 0$). Using formulas 13.4 and 13.2, we update the trend and the slope. The predicted sales are calculated as the product of the trend and the seasonal. The results are collected in Table 13.1.

13.7.3 Fitting the Model

With formulas 13.1–13.3, we calculate the three components of the Holt-Winters model and the predicted sales (column `pred sales`) for the rest of the training data, that is, from January 2014 to December 2016. Table 13.2 shows the results obtained for 2016. In the last row, the values of the trend and the slope will be used to forecast the sales of the first half of 2017. The seasonals of the upper half of the table (from January to June) will also be used.

Table 13.1 Holt-Winters model for 2013

Date	Sales	Trend (a)	Slope (b)	Seasonal (s)	Pred sales
2013–01	4,398	4,398.0	0.0	1.00	4,398.0
2013–02	4,887	4,495.8	19.6	1.00	4,495.8
2013–03	6,128	4,837.9	84.1	1.00	4,837.9
2013–04	6,137	5,165.0	132.7	1.00	5,165.0
2013–05	6,661	5,570.3	187.2	1.00	5,570.3
2013–06	6,418	5,889.6	213.6	1.00	5,889.6
2013–07	7,408	6,364.2	265.8	1.00	6,364.2
2013–08	8,122	6,928.4	325.5	1.00	6,928.4
2013–09	6,507	7,104.5	295.6	1.00	7,104.5
2013–10	7,634	7,446.9	305.0	1.00	7,446.9
2013–11	6,936	7,588.7	272.3	1.00	7,588.7
2013–12	6,422	7,573.2	214.8	1.00	7,573.2

Table 13.2 Holt-Winters model for 2016

Date	Sales	Trend (a)	Slope (b)	Seasonal (s)	Pred sales
2016–01	7,456	6,488.6	98.0	0.99	6,401.8
2016–02	6,424	6,656.9	112.1	0.93	6,216.0
2016–03	6,525	6,710.2	100.3	1.00	6,714.3
2016–04	6,466	6,729.5	84.1	1.00	6,727.8
2016–05	7,155	6,818.5	85.1	1.05	7,138.8
2016–06	6,238	6,736.1	51.6	1.01	6,788.9
2016–07	6,143	6,626.4	19.3	1.01	6,673.4
2016–08	6,674	6,678.5	25.9	0.98	6,571.1
2016–09	6,638	6,722.1	29.4	0.98	6,582.6
2016–10	6,705	6,737.7	26.7	1.00	6,749.5
2016–11	6,381	6,733.8	20.6	0.96	6,475.1
2016–12	6,859	6,772.9	24.3	1.00	6,799.7

Figure 13.2 is a line plot of the training data, with the Holt-Winters trend superimposed (dashed line). In the Excel file, this is the chart sheet `Line plot with HW trend (1)`. Please, note that the dashed line is the trend, not the predicted sales.

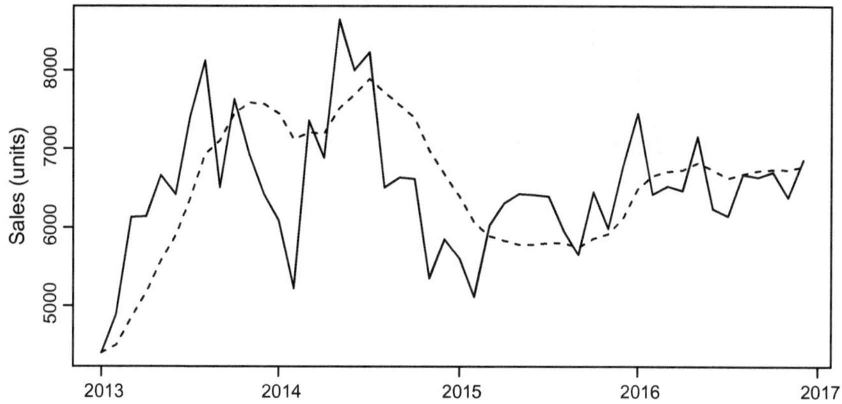

Fig. 13.2 Rib Eye sales with HW trend (alpha = 0.2)

13.7.4 Forecasting 6 Months Ahead

Finally, we calculate our forecast for the first half of 2017. The rationale is clear: we use the last slope and seasonal values available. The detail of the calculations is reported in Table 13.3. Note that the actual sales of the second column, included for comparison of the last column, are not involved in the calculation of the forecast.

13.7.5 Alternative Model

It may seem that we should update the components of the Holt-Winters more aggressively, given the sales variation shown in the line plot. We present here an alternative forecast, based on $\alpha = 0.5$, leaving the other two parameters as they were. Figure 13.3 shows the new trend. Compared to the trend of Fig. 13.2, this new trend captures more of the sales variation.

But it is not clear that what we are capturing is a trend and not a mix of seasonality and "noise", since the prediction errors are about the same. In Table 13.4, we report

Table 13.3 Holt-Winters forecast for 2017

Date	Sales	Trend (a)	Slope (b)	Seasonal (s)	Forecast
2017–01	7,768	6,797.2	24.3	0.99	6,706.2
2017–02	5,748	6,821.4	24.3	0.93	6,369.7
2017–03	7,617	6,845.7	24.3	1.00	6,849.8
2017–04	6,967	6,870.0	24.3	1.00	6,868.2
2017–05	7,081	6,894.2	24.3	1.05	7,218.1
2017–06	7,626	6,918.5	24.3	1.01	6,972.7

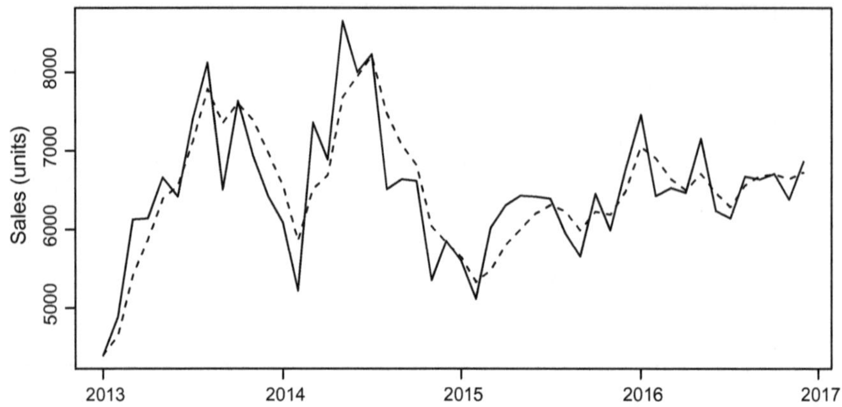

Fig. 13.3 Rib Eye sales with HW trend (alpha = 0.5)

Table 13.4 Alternative Holt-Winters forecast for 2017

Date	Sales	Trend (a)	Slope (b)	Seasonal (s)	Forecast
2017–01	7,768	6,752.0	28.4	1.00	6,760.0
2017–02	5,748	6,780.3	28.4	0.97	6,547.7
2017–03	7,617	6,808.7	28.4	1.03	7,005.0
2017–04	6,967	6,837.0	28.4	1.02	6,951.8
2017–05	7,081	6,865.4	28.4	1.04	7,145.8
2017–06	7,626	6,893.7	28.4	1.00	6,889.6

the forecasted sales for the first six months of 2017. The whole calculations are in the sheet `HW (alpha = 0.5)` of the Excel file.

13.8 Useful Tips

- With the Holt-Winters method, the forecasting model is updated as new information becomes available. Therefore, if the series is long enough, the impact of the initial estimates is negligible. Hence, *do not worry much about how to initialize the model in the first year.*
- When comparing two forecasting models, a visual inspection may not be enough to see clearly which is best. *In order to assess the accuracy of a model, calculate the mean absolute deviation or the standard deviation of the prediction error.*
- Forecasts are only valid when calculated for the immediate future. The Holt-Winters method is particularly useful for monthly or quarterly data. You just need to *adapt the formulas to the period you are concerned with, and update the forecast as soon as a new observation is available.*

R Code

A

A.1 Tata Auto Daily Returns

```
## Tata Auto Daily Returns ##

# Importing data #
url1 = 'https://raw.githubusercontent.com/quants-book/'
url2 = 'CSV_Files/master/tata.csv'
url = paste0(url1, url2)
df = read.csv(url, stringsAsFactors = FALSE)
str(df)
N = length(df$PRICE)

# Calculating the daily returns #
return = 100*(df$PRICE[-1]/df$PRICE[-N] - 1)

# Summary statistics #
mean(return)
sd(return)

# Calculating 95% limits #
lower = mean(return) - 2*sd(return)
lower
upper = mean(return) + 2*sd(return)
upper

# Counting the observations outside the limits #
sum(return < lower)
sum(return > upper)
mean(return < lower | return > upper)

# Getting the dates #
date = df$DATE[-1]
date[return < lower]
date[return > upper]
```


M. Á. Canela et al., *Quantitative Methods for Management*,
https://doi.org/10.1007/978-3-030-17554-2

```
# Figure 1.1 #
plot(return ~ as.Date(date),
    main = 'Figure 1.1. Tata Auto daily returns (line plot)',
    xlab = '', ylab = 'Daily returns (%)', type = 'l')

# Figure 1.2 #
hist(return,
    main = 'Figure 1.2. Tata Auto daily returns (histogram)',
    xlab = 'Daily returns (%)', ylab = 'Proportion',
    freq = FALSE)

# Figure 1.3 #
hist(return,
    main = 'Figure 1.3. Matching the normal distribution',
    xlab = 'Daily returns (%)', ylab = 'Proportion',
    freq = FALSE)
lines(seq(-6, 6, by=0.01), dnorm(seq(-6, 6, by=0.01),
    mean = mean(return), sd = sd(return)), lty = 4,
    lwd = 1.5)
```

A.2 The EuroLeague Final Four

```
## The EuroLeague Final Four ##

# Setting time locale to English (this may not be needed) #
Sys.setlocale('LC_TIME', 'English')

# Importing data #
url1 = 'https://raw.githubusercontent.com/quants-book/'
url2 = 'CSV_Files/master/euroleague'
url = paste0(url1, url2)
df_raw = read.csv(url, stringsAsFactors = FALSE)
str(df_raw)

# Raw data analysis #
mean(df_raw$count)
sd(df_raw$count)
median(df_raw$count)
quantile(df_raw$count, probs = c(0.25, 0.75))

# Figure 2.4 #
plot(count/10^3 ~ as.POSIXct(minute), data = df_raw,
    main = 'Figure 2.4. Tweets per minute (line plot)',
    xlab = '', ylab = 'Tweets (thousands)', type = 'l')

# Figure 2.5 #
hist(df_raw$count,
    main = 'Figure 2.5. Tweets per minute (histogram)',
    xlab = 'Tweets', ylab = 'Frequency')
```

```
# Aggregating data #
df_raw$hour = substr(df_raw$minute, 1, 13)
df_raw$hour = paste0(df_raw$hour, ':00:00')
df_agg = aggregate(count ~ hour, data = df_raw, sum)

# Aggregate data analysis #
mean(df_agg$count)
sd(df_agg$count)
median(df_agg$count)

# Figure 2.6 #
plot(count/10^3 ~ as.POSIXct(hour), data = df_agg,
     main = 'Figure 2.6. Tweets per hour (line plot)',
     xlab = '', ylab = 'Tweets (thousands)', type = 'l')

# Figure 2.7 #
hist(df_agg$count/10^3,
     main = 'Figure 2.7. Tweets per hour (histogram)',
     xlab = 'Tweets (thousands)', ylab = Frequency')

# Figure 2.8 #
plot(log(count) ~ as.POSIXct(hour), data = df_agg,
     main = 'Figure 2.8. Tweets per hour, log scale (line plot)',
     xlab = '', ylab = 'Tweets (log))', type = 'l')

# Figure 2.9 #
hist(log(df_agg$count),
     main = 'Figure 2.9. Tweets per hour, log scale (histogram)',
     xlab = 'Tweets (log)',nylab = 'Proportion', freq = FALSE)
lines(seq(4, 12, by=0.01), dnorm(seq(4, 12, by=0.01),
     mean = mean(log(df_agg$count)),
     sd = sd(log(df_agg$count))), lty = 4, lwd = 1.5)
```

A.3 Predicting Sales from Price

```
## Predicting Sales from Price ##

# Importing data #
url1 = 'https://raw.githubusercontent.com/quants-book/'
url2 = 'CSV_Files/master/greenchips.csv'
url = paste0(url1, url2)
df = read.csv(url)
str(df)

# Summary statistics #
c(mean(df$SALES), sd(df$SALES), cor(df$SALES, df$PRICE))
c(mean(df$PRICE), sd(df$PRICE))
```

```
# Regression line #
model = lm(formula = SALES ~ PRICE, data = df)
summary(model)

# Figure 3.2 #
plot(formula = SALES ~ PRICE, data = df,
   main = 'Figure 3.2.Regression line (R = -0.881)',
   xlab = 'Price (euros)', ylab = 'Sales (1000s)',
   pch = 16)
abline(coefficients(model))
```

A.4 Concrete Quality Control

```
## Concrete Quality Control ##

# Importing data #
url1 = 'https://raw.githubusercontent.com/quants-book/'
url2 = 'CSV_Files/master/concrete.csv'
url = paste0(url1, url2)
df = read.csv(url)
str(df)

# Regression line (1) #
mod1 = lm(formula = Resistance ~ Cement, data = df)
summary(mod1)

# Figure 4.1 #
plot(formula = Resistance ~ Cement, data = df,
    main = 'Figure 4.1. Resistance vs. cement (R = 0.786)',
    xlab = 'Cement (kg/m3)', ylab = 'Resistance (kg/cm2)',
    pch = 20)
abline(coefficients(mod1))

# Regression line (2) #
mod2 = lm(formula = Resistance ~ Additives, data = df)
summary(mod2)

# Figure 4.2 #
plot(formula = Resistance ~ Additives, data = df,
    main = 'Figure 4.2. Resistance vs. additives (R = 0.646)',
    xlab = 'Additives (kg/m3)', ylab = 'Resistance (kg/cm2)',
    pch = 20)
abline(coefficients(mod2))
```

```
# Regression line (3) #
mod3 = lm(formula = Resistance ~ Water, data = df)
summary(mod3)

# Figure 4.3 #
plot(formula = Resistance ~ Water, data = df,
    main = 'Figure 4.3. Resistance vs. water (R = 0.104)',
    xlab = 'Water (kg/m3)', ylab = 'Resistance (kg/cm2)',
    pch = 20)
abline(coefficients(mod3))

# Multiple regression analysis #
mod4 = lm(formula = Resistance ~ Cement + Additives +
    Water, data = df)
summary(mod4)

# Figure 4.4 #
plot(formula = Resistance ~ mod4$fitted.values, data = df,
    main = 'Figure 4.4. Resistance vs. predicted resistance',
    xlab = 'Predicted resistance (kg/cm2)',
    ylab = 'Resistance (kg/cm2)',
    pch = 20)
mod5 = lm(formula = Resistance ~ mod4$fitted.values,
    data = df)
abline(coefficients(mod5))
```

A.5 Orange Juice Pricing

```
## Orange Juice Pricing ##

# Importing data #
url1 = 'https://raw.githubusercontent.com/quants-book/'
url2 = 'CSV_Files/master/orange.csv'
url = paste0(url1, url2)
df = read.csv(url)
str(df)

# Regression line (1) #
mod1 = lm(formula = Mshare ~ MMaid, data = df)
summary(mod1)

# Multiple regression analysis (1) #
mod2 = lm(formula = Mshare ~ TropPremium + Trop + MMaid +
    Aldi, data = df)
summary(mod2)
```

```
# Correlation analysis #
round(cor(df[, -1]), 3)

# Multiple regression analysis (2) #
mod3 = lm(formula = Mshare ~ Trop + MMaid + Aldi,
    data = df)
summary(mod3)

# Multiple regression analysis (3) #
mod4 = lm(formula = Mshare ~ Trop + MMaid, data = df)
summary(mod4)
```

A.6 Gender Salary Gap

```
## Gender Salary Gap ##

# Importing data #
url1 = 'https://raw.githubusercontent.com/quants-book/'
url2 = 'CSV_Files/master/gender.csv'
url = paste0(url1, url2)
df = read.csv(url)
str(df)

# Creating dummy for male gender #
df$MALE = as.numeric(df$GENDER == 'Male')

# Gender salary gap #
tapply(df$SALARY, df$MALE, mean)

# Gender experience gap #
tapply(df$TENURE, df$MALE, mean)

# Regression line #
mod1 = lm(formula = SALARY ~ TENURE, data = df)
summary(mod1)

# Figure 6.1 #
plot(formula = SALARY ~ TENURE, data = df,
    main = 'Figure 6.1. Salary vs. tenure (R = 0.549)',
    xlab = 'Tenure (years)', ylab = 'Salary (US dollars)',
    pch = 20)
abline(coefficients(mod1))

# Regression analysis (1) #
mod2 = lm(formula = SALARY ~ MALE, data = df)
summary(mod2)
```

```
# Regression analysis (2) #
mod3 = lm(formula = SALARY ~ TENURE + MALE, data = df)
summary(mod3)
```

A.7 Diesel Consumption

```
## Diesel Consumption ##

# Importing data #
url1 = 'https://raw.githubusercontent.com/quants-book/'
url2 = 'CSV_Files/master/diesel.csv'
url = paste0(url1, url2)
df = read.csv(url)
str(df)

# Regression line (1) #
mod1 = lm(formula = EFFICIENCY ~ POWER, data = df)
summary(mod1)

# Figure 7.1 #
plot(formula = EFFICIENCY ~ POWER, data = df,
     main = 'Figure 7.1. Fuel efficiency vs. power use (R = -0.339)',
     xlab = 'Power use (%)', ylab = 'Fuel efficiency (km/l)',
     pch = 20)
abline(coefficients(mod1))

# Regression line (2) #
mod2 = lm(formula = EFFICIENCY ~ IDLE, data = df)
summary(mod2)

# Figure 7.2 #
plot(formula = EFFICIENCY ~ IDLE, data = df,
     main = 'Figure 7.2. Fuel efficiency vs. idle time (R = -0.257)',
     xlab = 'Idle time (%)', ylab = 'Fuel efficiency (km/l)',
     pch = 20)
abline(coefficients(mod2))

# Multiple regression analysis #
mod3 = lm(formula = EFFICIENCY ~ ., data = df)
summary(mod3)

# Splitting the sample #
f = function(x) c(mean(x), sd(x))
tapply(df$EFFICIENCY, df$DAIMLER, f)
tapply(df$IDLE, df$DAIMLER, f)
tapply(df$POWER, df$DAIMLER, f)
```

```
table(df$DAIMLER, df$YEARS)
mod4.1 = lm(formula = EFFICIENCY ~ IDLE + POWER + YEARS,
    data = df[df$DAIMLER==1,])
summary(mod4.1)
mod4.0 = lm(formula = EFFICIENCY ~ IDLE + POWER + YEARS,
    data = df[df$DAIMLER==0,])
summary(mod4.0)

# Analysis with interaction terms #
mod5 = lm(formula = EFFICIENCY ~ DAIMLER + IDLE +
    I(DAIMLER*IDLE) + POWER + I(DAIMLER*POWER) + YEARS +
    I(DAIMLER*YEARS), data = df)
summary(mod5)
```

A.8 Default at Alexia Bank

```
## Default at Alexia Bank ##

# Importing data #
url1 = 'https://raw.githubusercontent.com/quants-book/'
url2 = 'CSV_Files/master/alexia.csv'
url = paste0(url1, url2)
df = read.csv(url, stringsAsFactors = FALSE)
str(df)

# Data description #
table(df$GENDER, df$DEFAULT)
tapply(df$DEFAULT == 'Yes', df$GENDER, mean)
table(df$PART, df$DEFAULT)
tapply(df$DEFAULT == 'Yes', df$PART, mean)

# Regression analysis #
mod = lm(formula = (DEFAULT == 'Yes') ~ GENDER + AGE +
    CITIZEN + PART + INCOME + BALANCE, data = df)
summary(mod)

# Predictive scores #
scores = mod$fitted.values

# Confusion matrix (cutoff = 0.5) #
conf1 = table(scores > 0.5, df$DEFAULT == 'Yes')
conf1
acc1 = sum(diag(conf1))/sum(conf1)
acc1
tp1 = conf1['TRUE', 'TRUE']/sum(conf1[, 'TRUE'])
tp1
fp1 = conf1['TRUE', 'FALSE']/sum(conf1[, 'FALSE'])
fp1
```

```
# Confusion matrix (cutoff = 0.4) #
conf2 = table(scores > 0.4, df$DEFAULT == 'Yes')
conf2
acc2 = sum(diag(conf2))/sum(conf2)
acc2
tp2 = conf2['TRUE', 'TRUE']/sum(conf2[, 'TRUE'])
tp2
fp2 = conf2['TRUE', 'FALSE']/sum(conf2[, 'FALSE'])
fp2

# Confusion matrix (cutoff = 0.3) #
conf3 = table(scores > 0.3, df$DEFAULT == 'Yes')
conf3
acc3 = sum(diag(conf3))/sum(conf3)
acc3
tp3 = conf3['TRUE', 'TRUE']/sum(conf3[, 'TRUE'])
tp3
fp3 = conf3['TRUE', 'FALSE']/sum(conf3[, 'FALSE'])
fp3
```

A.9 The Churn Model

```
## The Churn Model ##

# Importing data #
url1 = 'https://raw.githubusercontent.com/quants-book/'
url2 = 'CSV_Files/master/churn.csv'
url = paste0(url1, url2)
df = read.csv(url, stringsAsFactors = FALSE)
str(df)

# Churning rate #
sum(df$CHURN)

# Crosstabulation of data plan variables #
table(df$DATAPLAN, df$DATAGB)

# Splitting the data set #
train = sample(1:5000, size = 2500, replace = FALSE)
df_train = df[train, ]
df_test = df[-train, ]

# Regression equation #
fm = CHURN ~ ACLENGTH + INTPLAN + DATAPLAN + OMMIN +
    OMCALL + OTMIN + OTCALL + NGMIN + NGCALL +
    IMIN + ICALL + CUSCALL
mod = lm(formula = fm, data = df_train)
summary(mod)
```

```
# Evaluation in the training set (cutoff = 0.19) #
pred_train = predict(mod, newdata = df_train)
conf_train = table(pred_train > 0.19, df_train$CHURN)
conf_train
acc_train = sum(diag(conf_train))/sum(conf_train)
tp_train = conf_train['TRUE', '1']/sum(conf_train[, '1'])
tp_train
fp_train = conf_train['TRUE', '0']/sum(conf_train[, '0'])
fp_train

# Evaluation in the test set(cutoff = 0.19) #
pred_test = predict(mod, newdata = df_test)
conf_test = table(pred_test > 0.19, df_test$CHURN)
conf_test
acc_test = sum(diag(conf_test))/sum(conf_test)
acc_test
tp_test = conf_test['TRUE', '1']/sum(conf_test[, '1'])
tp_test
fp_test = conf_test['TRUE', '0']/sum(conf_test[, '0'])
fp_test
```

A.10 Polar Bear Sales

```
## Polar Bear Sales ##

# Importing data #
url1 = 'https://raw.githubusercontent.com/quants-book/'
url2 = 'CSV_Files/master/polar.csv'
url = paste0(url1, url2)
df = read.csv(url, stringsAsFactors = FALSE)
str(df)

# Defining ts object #
xt = ts(data = df$SALES, start = c(2011, 1),
    end = c(2016, 12), frequency = 12, deltat = 1/12)

# Figure 10.1 #
plot(xt, main = 'Figure 10.1. Polar Bear monthly sales',
    xlab = '', ylab = 'Sales (units)')

# Fitting a linear trend #
N = nrow(df)
t = 1:N
trend = lm(xt ~ t)
summary(trend)
at = trend$fitted.values
```

```
# Figure 10.2 #
plot(xt,
    main = 'Figure 10.2. Polar Bear data with linear trend',
    xlab = '', ylab = 'Sales (units)')
lines(2011 + (t-1)/12, at, lty = 2)

# Calculating the seasonal factors #
month = substr(df$MONTH, 6, 7) #
st = tapply(xt/at, month, mean)

# Figure 10.3 #
plot(st, type = 'b', pch = 16,
    main = 'Figure 10.3. Seasonal factors',
    xlab = 'Month', ylab = '')

# Figure 10.4 #
st = rep(st, 6)
plot(xt, main = 'Figure 10.4. Predicted vs. actual sales',
    xlab = '', ylab = 'Sales (units)')
lines(2011 + (t-1)/12, at*st, lty = 2)

# Prediction error #
zt = ts(xt - at*st, start = c(2011, 1), end = c(2016, 12),
    frequency = 12, deltat = 1/12)

# Figure 10.5 #
plot(zt, main = 'Figure 10.5. Prediction error',
    xlab = '', ylab = 'Sales (units)')

# Forecasting year 2017 #
a1 = trend$coefficients[1]
a2 = trend$coefficients[2]
ft = (a1 + a2*(73:84))*st[1:12]

# Table 10.1 #
tab1 = c('JAN', 'FEB', 'MAR', 'APR', 'MAY', 'JUN', 'JUL', 'AUG',
    'SEP', 'OCT', 'NOV', 'DEC')
tab2 = round(a1 + a2*(73:84), 1)
tab3 = round(st[1:12], 2)
tab4 = round(ft, 1)
cbind(tab1, tab2, tab3, tab4)
```

A.11 Bayou Beer Sales

```
## Bayou Beer Sales ##

# Importing data #
url1 = 'https://raw.githubusercontent.com/quants-book/'
url2 = 'CSV_Files/master/bayou.csv'
url = paste0(url1, url2)
df = read.csv(url, stringsAsFactors = FALSE)
str(df)

# Figure 11.1 #
xt = ts(data = df$SALES, start = c(2008, 1),
    end = c(2017, 12), frequency = 12, deltat = 1/12)
plot(xt, main ='Figure 11.1 Bayou Beer sales',
    xlab = '', ylab = 'Thousand barrels')

# Splitting the data set #
df1 = df[df$YEAR < 2017, ]
df2 = df[df$YEAR == 2017, ]

# Fitting a linear trend (2008-2016) #
t = 1:108
lmod = lm(formula = SALES ~ t, data = df1)
summary(lmod)
ltrend = lmod$fitted.values

# Fitting a quadratic trend (2008-2016) #
t2 = t^2
qmod = lm(formula = SALES ~ t + t2, data = df1)
summary(qmod)
qtrend = qmod$fitted.values

# Figure 11.2 #
plot(xt,
    main = 'Figure 11.2. Bayou Beer sales with linear
            and quadratic trends',
    xlab = '', ylab = 'Thousand barrels')
lines(2008 + (t - 1)/12, ltrend)
lines(2008 + (t - 1)/12, qtrend, lty = 2)

# Splitting the data set #
df1 = df[df$YEAR < 2017, ]
df2 = df[df$YEAR == 2017, ]

# Calculating the seasonals #
month = rep(1:12, 9)
seasonal = tapply(df1$SALES - qtrend, month, mean)

# Figure 11.3 #
plot(seasonal, main = 'Figure 11.3. Seasonals',
    xlab = 'Month', ylab = 'Thousand barrels', type = 'b',
    pch = 16)
```

```
# Predicted sales and prediction error #
pred_sales = qtrend + rep(seasonal, 9)
pred_error = df1$SALES - pred_sales

# Figure 11.4 #
plot(xt, main = 'Figure 11.4. Predicted vs. actual sales',
    xlab = '', ylab = 'Thousand barrels')
lines(2008 + (t - 1)/12, pred_sales, lty = 2)

# Figure 11.5 #
et = ts(data = pred_error, start = c(2008, 1),
    end = c(2017, 12), frequency = 12, deltat = 1/12)
plot(et, main = 'Figure 11.5. Prediction error',
    xlab = '', ylab = 'Thousand barrels')

# Forecasting year 2017 #
b0 = qmod$coefficients[1]
b1 = qmod$coefficients[2]
b2 = qmod$coefficients[3]
forecast = b0 + b1*(109:120) + b2*(109:120)^2 + seasonal

# Table 11.1 #
tab1 = c('JAN', 'FEB', 'MAR', 'APR', 'MAY', 'JUN', 'JUL',
    'AUG', 'SEP', 'OCT', 'NOV', 'DEC')
tab2 = round(b0 + b1*(109:120) + b2*(109:120)^2, 3)
tab3 = round(seasonal, 3)
tab4 = round(forecast, 3)
tab5 = round(df2$SALES, 3)
tab6 = round(df2$SALES - forecast, 3)
cbind(tab1, tab2, tab3, tab4, tab5, tab6)
```

A.12 Brandy Consumption in Australia

```
## Brandy Consumption in Australia ##

# Importing data #
url1 = 'https://raw.githubusercontent.com/quants-book/'
url2 = 'CSV_Files/master/brandy.csv'
url = paste0(url1, url2)
df = read.csv(url, stringsAsFactors = FALSE)
str(df)

# Figure 12.3 #
xt = ts(data = df$SALES, start = c(1985, 1),
    end = c(2014, 2), frequency = 4, deltat = 1/4)
plot(xt,
    main = 'Figure 12.3. Brandy quarterly sales',
    xlab = '', ylab = 'Brandy sales (1000 litres alcohol)')
```

```
# Dropping year 2014 #
df = df[1:(nrow(df) - 2), ]

# Figure 12.4 #
xt = ts(data = df$SALES, start = c(1985, 1),
    end=c(2013,4), frequency = 4, deltat = 1/4)
N = nrow(df)
t = 1:N
lt = lm(xt ~ t)$fitted.values
qt = lm(xt ~ t + I(t^2))$fitted.values
plot(xt,
    main = 'Figure 12.4. Brandy sales with linear and quadratic
            trends',
    xlab = '', ylab = 'Brandy sales (1000 litres alcohol)')
lines(1985 + (t - 1)/4, lt)
lines(1985 + (t - 1)/4, qt, lty = 2)

# Figure 12.5 #
mat = (1/8)*xt[1:(N - 4)] + (1/4)*xt[2:(N - 3)] +
    (1/4)*xt[3:(N - 2)] + (1/4)*xt[4:(N - 1)] + (1/8)*xt[5:N]
mat = c(NA, NA, mat, NA, NA)
plot(xt, main = 'Figure 12.5. Brandy sales with MA trend',
    xlab = '', ylab = 'Brandy sales (1000 litres alcohol)')
lines(c(NA, NA, 1985 + (t[2:(N - 3)])/4, NA, NA), mat, lty = 2)

# Dropping years 1985-99 #
df = df[65:N, ]
xt = ts(data = df$SALES, start = c(2001, 1),
    end = c(2013, 4), frequency = 4, deltat = 1/4)
qt = qt[65:N]

# Figure 12.6 #
quarter = substr(df$DATE, 6, 7)
seasonals = tapply(xt/qt, quarter, mean)
st = rep(seasonals, 13)
dt = xt/st
plot(xt, main = 'Figure 12.6. Deseasonalized brandy sales',
    xlab = '', ylab = 'Brandy sales (1000 litres alcohol)')
t = 1:52
lines(2001 + (t - 1)/4, dt, lty = 2)

# Figure 12.7 #
et = dt[1]
for(i in 2:52) et[i] = 0.2*dt[i] + 0.8*et[i - 1]
plot(xt,
    main = 'Figure 12.7. Brandy sales with exponential smoothing
            trend',
    xlab = '', ylab = 'Brandy sales (1000 litres alcohol)')
lines(2001 + (t - 1)/4, et, lty = 2)
```

A.13 Supermercados Andinos

```
## Supermercados Andinos ##

# Importing data #
url1 = 'https://raw.githubusercontent.com/quants-book/'
url2 = 'CSV_Files/master/andinos.csv'
url = paste0(url1, url2)
df = read.csv(url, stringsAsFactors = FALSE)
str(df)

# Figure 13.1 #
xt = ts(data = df$sales, start = c(2013, 1),
     end = c(2017, 6), frequency = 12, deltat = 1/12)
plot(xt, main = 'Figure 13.1. Rib Eye monthly sales',
     xlab = '', ylab = 'Sales (units)')

# Holt-Winters approximation (1) #
alpha = 0.2
beta = 0.2
gamma = 0.2
xt = ts(data = xt[1:48], start = c(2013, 1),
     end = c(2016, 12), frequency = 12, deltat = 1/12)
st = rep(1, 12)
at = xt[1]/st[1]
bt = 0
for(i in 2:12) {
     at[i] = alpha*xt[i]/st[i] +
          (1 - alpha)*(at[i - 1] + bt[i - 1])
     bt[i] = beta*(at[i] - at[i - 1]) + (1 - beta)*bt[i - 1]
}
for(i in 13:48) {
     at[i] = alpha*(xt[i]/st[i - 12]) +
          (1 - alpha)*(at[i - 1] + bt[i - 1])
     bt[i] = beta*(at[i] - at[i - 1]) + (1 - beta)*bt[i - 1]
     st[i] = gamma*(xt[i]/at[i]) + (1 - gamma)*st[i - 12]
}
hwt = at*st

# Figure 13.2 #
plot(xt,
     main = 'Figure 13.2. Rib Eye sales with HW trend (alpha = 0.2)',
     xlab = '', ylab = 'Sales (units)')
t = 1:48
lines(2013 + (t-1)/12, at, lty = 2)

# Holt-Winters forecasting (1) #
at[49:54] = at[48] + (1:6)*bt[48]
hwt[49:54] = at[49:54]*st[37:42]
```

```
# Holt-Winters approximation (2) #
alpha = 0.5
beta = 0.2
gamma = 0.2
xt = ts(data = xt[1:48], start = c(2013, 1),
     end = c(2016, 12), frequency = 12, deltat = 1/12)
st = rep(1, 12)
at = xt[1]/st[1]
bt = 0
for(i in 2:12) {
     at[i] = alpha*xt[i]/st[i] +
          (1 - alpha)*(at[i - 1] + bt[i - 1])
     bt[i] = beta*(at[i] - at[i - 1]) + (1 - beta)*bt[i - 1]
}
for(i in 13:48) {
     at[i] = alpha*(xt[i]/st[i - 12]) +
          (1 - alpha)*(at[i - 1] + bt[i - 1])
     bt[i] = beta*(at[i] - at[i - 1]) + (1 - beta)*bt[i - 1]
     st[i] = gamma*(xt[i]/at[i]) + (1 - gamma)*st[i - 12]
}
hwt = at*st
# Figure 13.3 #
plot(xt,
     main = 'Figure 13.3. Rib Eye sales with HW trend (alpha = 0.5)',
     xlab = ", ylab = 'Sales (units)')
t = 1:48
lines(2013 + (t-1)/12, at, lty = 2)

# Holt-Winters forecasting (2) #
at[49:54] = at[48] + (1:6)*bt[48]
hwt[49:54] = at[49:54]*st[37:42]
```